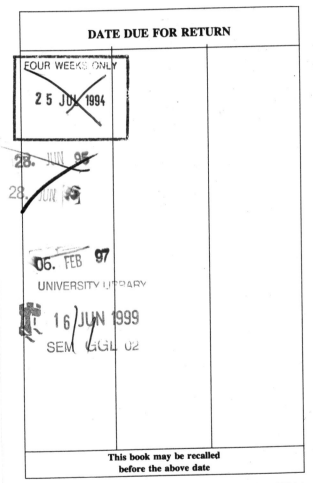

ORGANIC SYNTHESIS
THEORY AND APPLICATIONS

Volume 1 · 1989

ORGANIC SYNTHESIS
THEORY AND APPLICATIONS

A Research Annual

Editor: TOMAS HUDLICKY
Department of Chemistry
Virginia Polytechnic Institute
and State University

VOLUME 1 · 1989

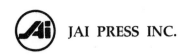 JAI PRESS INC.

Greenwich, Connecticut   *London, England*

CONTENTS

INTRODUCTION TO THE SERIES:
AN EDITOR'S FOREWORD

The field of organic chemistry has developed dramatically during the past forty years. Thus it appears to be an opportune time to publish a series of essays on various relevant themes in the 80s written by workers who are active in the discipline. This collection includes many of the important areas of current research interest. To cover such a broad area a very substantial effort is needed, as was the cooperation of a large number of colleagues and friends who have agreed to act as series editors. I have been gratified by the favorable response of research workers in the field to the invitation to contribute chapters in their own specialities. Each contributor has writen a critical, lively and up-to-date description of his field of interest and competence, so that the chapters are not merely literature surveys. It is hoped that this new and continuing series will prove valuable to active researchers, and that many new ideas will be generated for future theoretical and experimental research. The wide coverage of material should be of interest to graduate students, postdoctoral fellows and those teaching specialized topics to graduate students.

Department of Chemistry   *Albert Padwa*
Emory University   *Consulting Editor*
Atlanta, Georgia

PREFACE

The series entitled *Organic Synthesis: Theory and Applications* will provide a summary of the state of the art synthetic methodology of the 1980s. This first volume is intended as an introduction to some of the topics that enjoyed widespread popularity during the past decade. The developments associated with, for example, the use of the Diels–Alder reaction are truly astonishing and leave only fine details to be worked out in this field. Similarly, the carbocyclic silane-mediated cyclizations are at the point today where precise physical organic parameters need to be identified and established to render these processes permanent and reliable tools of the synthetic chemist. The chapter on nonconventional methods of synthesis mirrors the trend of the last 15 years in which organic chemists ventured to other areas of science in search of novel means of energizing the reactions pathways.

The development of organic synthesis in the post-Woodward years (1960–present) has traveled a path that diverged from the total synthesis focus of the 1960s on to the details of the processes involved in total synthesis. Indeed while an astonishing amount of progress has been reached in the area of especially the enantioselective methodology, very little has changed in the way of design of complex molecules other than what the Woodward legacy left us. The daily use of MM2 calculation by the organic community is perhaps the only improvement in the way molecular architects guide their approach to complex targets. Furthermore, the advances in such procedures as aldol condensation and Diels–Alder cycloaddition still fall short of achieving the fundamental ability of chemists to predict as well as explain the pathways of organic reaction.

While every trained chemist can readily justify the outcome of a known experiment using the principles of mechanistic organic chemistry, the same chemist with a pencil, a chalk, or a computer terminal runs a little better than 50% chance of predicting the outcome of a reaction yet to be performed. Clearly then, the theoretical foundation that we use is inadequate and incomplete if such discrepancies are possible. What is the solution? Amendments or complete overhaul of mechanistic rationalizations in the days to come to achieve better powers of prediction.

The next volumes will examine some of the mathematical concepts emerging in organic synthesis today as well as some applications of novel design elements to the synthesis of complex molecules. The bioorganic and polymer subsets of our discipline are increasingly making their influence on the preparation and management of organic compounds. The next decade will undoubtedly witness a marriage between all of these disciplines as greater understanding of the fundamentals of organic chemistry will bring about the much needed departure from the way we currently do business. It is hoped that this first volume will start the reader thinking about such changes.

Tomas Hudlicky
Series Editor

ASYMMETRIC DIELS–ALDER REACTIONS

Michael J. Taschner

OUTLINE

Organic Synthesis: Theory and Application, Vol. 1, pages 1–101.
Copyright © 1989 by JAI Press Inc.
All rights of reproduction in any form reserved.
ISBN: 0-89232-865-7

I. INTRODUCTION

The Diels–Alder reaction has evolved into one of the most effective weapons in the arsenal of the organic chemist since the first such reaction was reported by Diels and Alder in 1928.[1] Its remarkable regioselectivity,[2] relatively predictable endo selectivity,[3] syn stereospecificity,[4] and capacity to control the relative stereochemistry at up to four of the newly created chiral centers render the Diels–Alder reaction unequalled in terms of elegance and efficiency in the construction of six-membered rings. As organic synthesis matured, the ability to control the relative stereochemistry was no longer sufficient. Organic chemists had become increasingly concerned with the control of absolute stereochemistry.[5] Therefore; it was only natural that during the course of this development, both the intermolecular Diels–Alder reaction[6] and its intramolecular counterpart[7] would be proded and probed with the hope they would be capable of inducing asymmetry in the bond-forming process in a controllable and predictable manner.[8]

The development of the strategies to attain absolute stereochemical control in the intermolecular and intramolecular [4 + 2] cycloaddition processes will be presented within the individual sections in as chronological an order as coherently possible. When appropriate, mechanistic interpretations concerning potential diastereomeric transition states of a particular transformation will be advanced in an effort to furnish a better understanding and appreciation of the overall process. The application of particular tactics in the context of enantioselective total syntheses of a number of natural and a few unnatural products will be interspersed within the sections. An effort was made to be as thorough as possible in the coverage of the pertinent literature to date. However, it is inevitable that a number of articles will have been inadvertently and unintentionally omitted. Apologies are extended in advance to the authors of any of the articles that may have been overlooked.

II. CHIRAL DIENOPHILES

A preliminary communication from Walborsky in 1961 reported the inaugural investigation[9] directed at the induction of asymmetry in the Diels–Alder process.[10] The reaction between (−)-dimenthyl fumarate 1 and 1,3-butadiene under thermal conditions produced the cyclohexenyl diester 2. In order to avoid any possibility for resolution, the diester was immediately reduced with LiAlH$_4$ to afford diol 3 with the (1R,2R)-configuration in 2.4% enantiomeric excess (ee). The absolute configuration was established by chemical conversion of 3 to (−)-threo-3,4-dimethyladipic acid (4), whose absolute configuration was known. In the presence of AlCl$_3$, the reaction of

1 and butadiene produced the same product **2**, but was of opposite sign and configuration. Reduction of this diester with LiAlH$_4$ yielded diol **3** in 57% ee which possessed the (1*S*,2*S*)-configuration.

Two years later, Walborsky published the full details of the study that showed the results to be dependent on solvent, temperature, and catalyst. In the thermal reaction, the optical purity of the (*R*)-(−)-isomer was found to increase with increasing temperature (3% ee at 180°C). However, this minimal increase in optical yield was more than offset by the decrease in chemical yield. The best results were obtained using Lewis acid conditions with 1 equivalent of TiCl$_4$ in toluene at 25°C. After reduction of the diester, diol **3** was isolated in 78% optical purity and 80% yield. The results have been explained by means of a modified Prelog model in which the conformation of the fumarate resembles that shown in **5**.[11]

Although the cycloadditions of **1** were dependent on solvent and temperature, their effect was minimal. The use of Lewis acid catalysis greatly
enhanced the induction, but the adducts were of opposite absolute configuration to those obtained under thermal conditions. This prompted Jurczak to
investigate the use of high pressure in this asymmetric Diels–Alder reaction.
There was a modest increase in the diastereoselectivity (12.8%); however,
the absolute configuration was the same as that observed in the Lewis acid-
catalyzed reactions and remained unchanged over the range of pressures
examined.[12]

In the quest for new alcohol reagents capable of imparting some diastereofacial differentiation, the mixed fumarate esters of the isoborneols **6** and **7**
were prepared. The reactions with anthracene as the diene were examined to
determine the efficiency of these dienophiles to induce asymmetry in the
Diels–Alder reaction. As evidenced in Table 1, the uncatalyzed reaction of
6c with anthracene and subsequent $LiAlH_4$ reduction results in the formation of the (11S,12S)-diastereomer **8** in 90% yield [60% diastereomeric
excess (de)]. The uncatalyzed reaction of **7c** produces the adduct with the
expected opposite (11R,12R)-configuration, but in reduced yield (68%) and
diastereoselectivity (34%). In combination with $AlCl_3$, the dienophiles
exhibit increased reactivity and better diastereofacial selectivity. The reaction of **6a** in the presence of $AlCl_3$ provides the (11S,12S)-diastereomer in
96% yield with a diastereomeric excess of 99%. The Lewis acid-mediated
reaction of **7c** affords a quantitative yield of the (11R,12R)-diastereomer
with a diastereomeric excess of 92%.[13]

a, R= CH_2Ph
b, R= $CHPh_2$
c, R= $CONHPh$

(11S, 12S)-**8**   (11R, 12R)-**8**

The catalyzed and uncatalyzed reactions of mono- and dibornyl fumar-
ates with diphenylisobenzofuran have been used in an analysis of the
transition state geometry of the cycloaddition. Tolbert found the amount of
asymmetric induction provides a useful probe of transition state geometry.

Table 1.   Cycloadditions of **6** and **7** with Anthracene

Dienophile	Catalyst	Temperature (°C)	% Yield	% de
6a	—	110	30	< 10
6b	—	110	68	< 10
6c	—	110	90	60
7c	—	110	68	34
6a	AlCl$_3$	−30	96	> 99
6c	AlCl$_3$	−30	100	99
7c	AlCl$_3$	0	100	92

He concluded the Diels–Alder reaction was synchronous only when a catalyst was not available to polarize the transition state and the effect of the Lewis acid in enhancing asymmetric induction was to increase the steric interaction at the end furthest away from the complexed catalyst.[14]

Two groups have reported the cycloadditions of chiral acrylates **9a–9e** with cyclopentadiene. The results were consistent with the Prelog model hypothesized by Walborsky. However, as can be seen in Table 2, the enantiomeric excesses for the uncatalyzed reactions were low, and those for the Lewis acid-catalyzed processes were moderate to good.[15]

9

a, R= (R)-(-)-menthyl
b, R= (R)-(-)-2-octyl
c, R= (S)-(+)-2-octyl
d, R= (S)-(+)-2-(3,3-dimethyl)butyl
e, R= (S)-(+)-2-butyl

Table 2.   Reactions of Acrylates **9** with Cyclopentadiene

Acrylate	Temperature (°C)	Catalyst	**10** (% ee)		**11** (% ee)		Reference
9a	0	—	(+)	9.1	—	—	15a
9a	−70	AlCl$_3$·Et$_2$O	(+)	67	—	—	15a
9a	−70	BF$_3$·Et$_2$O	(+)	82	—	—	15a
9a	4	SnCl$_4$	(+)	41	—	—	15b
9b	35	—	(+)	4.1	(+)	3	15a
9b	−70	BF$_3$·Et$_2$O	(+)	27	—	—	15a
9c	35	—	(−)	4.1	—	—	15a
9c	−70	BF$_3$·Et$_2$O	(−)	28	—	—	15a
9c	4	SnCl$_4$	(−)	15	—	—	15b
9d	35	—	(−)	11	—	—	15a
9d	−70	BF$_3$·Et$_2$O	(−)	88	—	—	15a
9e	4	SnCl$_4$	(−)	24	—	—	15b

An isolated study of acrylate **9a** with 1,3-cyclohexadiene under thermal conditions was reported to produce predominantly the (R)-(+)-*endo*-isomer **12** in 60% ee after LiAlH$_4$ reduction.[16]

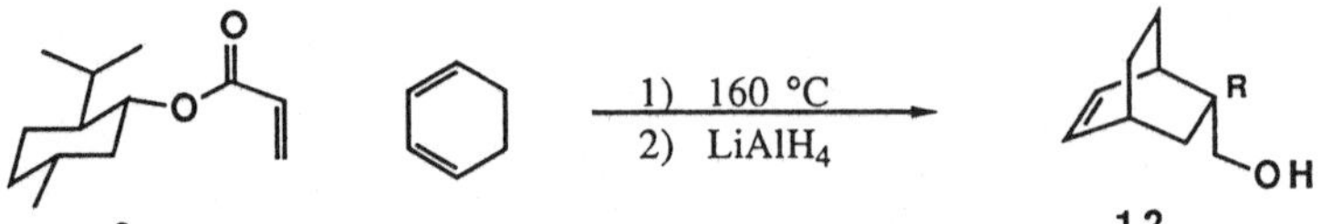

The turning point in the development of asymmetric induction in the Diels–Alder reaction came when Corey and Ensley reported the preparation of an optically active intermediate for the synthesis of prostaglandins via an asymmetric Diels–Alder reaction, which utilized the 8-phenylmenthyl auxiliary in combination with the acrylate dienophile.[17] They found the 8-phenylmenthyl group to be more effective at differentiating between the diastereotopic faces of the acrylate. The SnCl$_4$-catalyzed reaction of **13** with cyclopentadiene and subsequent LiAlH$_4$ reduction was originally reported to provide (+)-**10** in 99% ee. A reinvestigation of this reaction by Oppolzer corrected the value to 89% ee.[18] The improvement in diastereoselection has been attributed to the reactive conformation being similar to the one shown below. The acrylate is assumed to have the antiplanar arrangement of the olefin and the carbonyl which results in the shielding of the *re*-face of the acrylate by the phenyl substituent, as well as allowing for π-stacking interactions between the acrylate and the aromatic ring. It is believed the combination of these steric and electronic effects are responsible for the heightened selectivity.

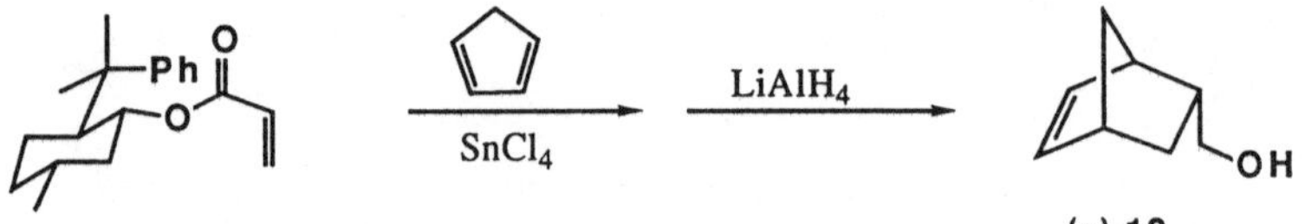

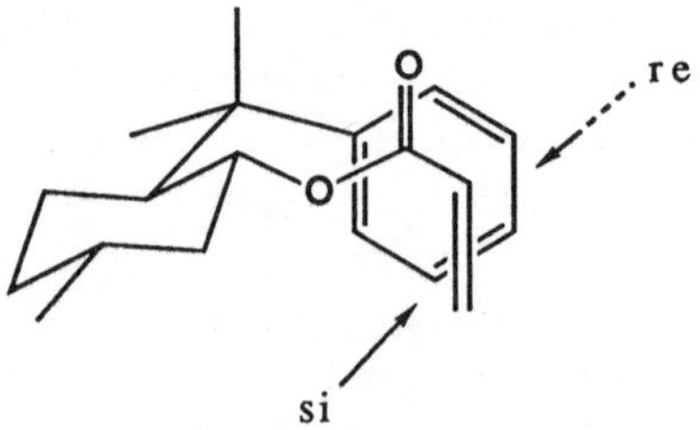

For the prostaglandin work, **13** was reacted with cyclopentadiene **14** in the presence of AlCl$_3$ to furnish an 89% yield of the (−)-*endo*-adduct **15**. This was transformed by a series of reactions into the known, optically active iodolactone **16**, which had previously been employed in the preparation of a number of prostaglandins. The preparation of **16** also proved the absolute configuration of **15** is as shown.[17]

Green used the $BF_3 \cdot Et_2O$-catalyzed cycloaddition of the enantiomer of **9d** with cyclopentadiene in a total synthesis of (+)-brefeldin-A (**19**). The adduct **17**, obtained in 75% yield, was transformed into norbornenone (−)-**18** (80–85% ee). This was converted into (+)-**19** by previously established procedures.[19]

In 1980, Boeckman reported an efficient enantioselective synthesis of the antitumor agent sarkomycin from the 8-phenylmenthyl acrylate **13**.[20] Walborsky's model predicted the cycloaddition of **13** with butadiene would produce the required (*R*)-configuration in the cyclohexenyl ester for the ultimate conversion to (*R*)-(−)-sarkomycin (**22**). The reaction between acrylate **13** and butadiene in the presence of $TiCl_4$ produced the desired cycloadduct. Other Lewis acids were reported to be inferior in terms of chemical or optical yields. The ester was then reduced with $LiAlH_4$ to provide (*R*)-(+)-**20** in 70% chemical yield and 86–91% ee. The alcohol was converted to the bicyclic keto lactone **21**, which was further transmuted in a series of reactions to (*R*)-(−)-**22**.

The degree of asymmetric induction in all the Diels–Alder reactions of **9a** and **13** prior to 1981 was determined entirely by chiroptic measurements of the LiAlH$_4$ product **10**. Because this method was not reliable, Oppolzer undertook a reinvestigation of the Diels–Alder reaction of these two dienophiles under a variety of conditions. The extent of asymmetric induction was measured using ^{19}F-NMR spectroscopy of the Mosher esters derived from alcohol **10**. This method was found to be a more reliable technique for the determination of the enantiomeric purity of the products as well as a useful check of the enantiomeric purity of the chiral auxiliary. Shown in Table 3 is a summary of the results obtained in the systematic studies of the reactions of **9a** and **13** in which the solvent, temperature, and Lewis acid were varied.[18]

Table 3. Systematic Study of the Cycloadditions of **9a** and **13**

Acrylate	Equivalent Lewis acid	Solvent	Temperature (°C)	Time (h)	% Yield	endo/exo	% ee of 2-(R)-10
9a	1.0 SnCl$_4$	PhCH$_3$	0	0.5	—	—	51
13	1.5 SnCl$_4$	PhCH$_3$	0	3.5	95	84 : 16	89
9a	0.7 AlCl$_3$	CH$_2$Cl$_{23}$	−55	—	—	—	48
13	0.7 AlCl$_3$	CH$_2$Cl$_2$	−20	3.5	89	91 : 9	65
13	0.7 AlCl$_3$	PhCH$_3$	−20	3.5	96	92 : 8	52
9a	1.5 Me$_2$AlCl	CH$_2$Cl$_2$	0	3.5	73	92 : 8	47
13	1.5 Me$_2$AlCl	CH$_2$Cl$_2$	0	3.5	95	89 : 11	64
13	1.5 Me$_2$AlCl	PhCH$_3$	0	3.5	81	88 : 12	55
9a	1.5 TiCl$_4$	CH$_2$Cl$_2$	−20	3.5	65	92 : 8	62
13	1.5 TiCl$_4$	CH$_2$Cl$_2$	−20	3.5	83	89 : 11	90

The table shows that the most useful conditions involve the use of either 1.5 equivalents of SnCl$_4$ in toluene at 0°C (89% ee) or 1.5 equivalents of TiCl$_4$ in CH$_2$Cl$_2$ (90% ee). Under all conditions studied, the predominant product, as demonstrated previously, proved to be the 2-(R)-enantiomer. Also corroborating previous work, the 8-phenylmenthyl group was found to be superior to the menthyl group for the induction of chirality into the cycloadducts.

Table 4. Cycloadditions of Acrylates Derived from **23–29** with Cyclopentadiene

Starting material	Alcohol	Adduct configuration	% ee
(+)-Pulegone	**23**	R	88
(+)-Limonene	**24**	R	63
(−)-β-Pinene	**25**	R	85
(+)-4-Cholesten-3-one	**26**	S	84
(+)-Camphor	**27a**, R_1 = Ph; R_2 = H	S	88
(+)-Camphor	**28**	R	81.5
(−)-Camphor	**29**	S	82.7

Although the chiral inductions with the 8-phenylmenthyl group were very good, there were still some shortcomings with the overall implementation of the asymmetric Diels–Alder strategy using this auxiliary. The two major limitations are the need to purify the $(-)$-8-phenylmenthol by careful medium pressure chromatography and the relative inaccessibility of its *re*-face directing optical antipode. To overcome these problems, Oppolzer initiated a program designed to prepare more effective and versatile chiral auxiliaries. Utilizing the insight obtained from the 8-phenylmenthyl studies concerning the shielding of one acrylate face via the π-stacking interactions, investigations concentrated on the preparation and use of the *si*-face and *re* face directing auxiliaries **23–29**. These were transformed into their acrylate esters and then condensed with cyclopentadiene in the presence of TiCl$_4$ to afford the cycloadducts. The adducts were reduced to either $(+)$-(R)-**10** or $(-)$-(S)-**10** with LiAlH$_4$ for the determination of chiral induction. Shown in Table 4 are the results of this study.[21]

Examination of the results reveals a number of important findings. The acrylates derived from alcohols **26**, **27a**, and **29** were the first readily available *re*-face directing acrylates developed which gave excellent chiral inductions. The alcohols **27a** and **29** have the additional advantages in that both are crystalline and both enenatiomeric forms of the precursors are readily available. Also worth noting is the result from alcohol **25**. This produces an absolute configuration and enantiomeric excess comparable to that obtained with **13**. It also is crystalline and both enantiomeric precursors are readily available.

Encouraged by the results obtained with alcohol **27a**, the acrylates derived from the *cis*-3-hydroxyisobornyl ethers **27**, **30**, and **31** were prepared to examine how important a role the hypothetical aryl/acrylate π-stacking played in the asymmetric Diels–Alder reaction with cyclopentadiene (Table 5). With most of these acrylates, TiCl$_4$ proved to be too harsh a Lewis acid and usually caused rapid ether cleavage. This problem was overcome by the use of the milder Lewis acid TiCl$_2$(Oi-Pr)$_2$ [TiCl$_4$/Ti(Oi-Pr)$_4$, 1:1]. The milder conditions afforded the adducts in excellent chemical yields and produced diastereomeric excesses of up to 92%.

27   **30**   **31**

In light of these results, there appears to be no particular advantage to increasing the aromatic surface of the appendant ether. This seems to imply

Table 5.   TiCl$_2$(Oi-Pr)$_2$ Catalyzed Reactions of Acrylates Derived from **27**, **30**, and **31** with Cyclopentadiene

Alcohol	R$_1$	R$_2$	Temperature (°C)	Yield	endo/exo	endo configuration	% de
27a	(phenyl)	H	0	91	86 : 14	S	46
27b	(phenyl)	(phenyl)	−20	94	90 : 10	S	72
27c	(naphthyl)	H	−20	98	90 : 10	S	69
27d	(naphthyl)	H	0	97	85 : 15	S	54
27e	t-Bu	H	−20	95	96 : 4	S	97
30a	(phenyl)	(phenyl)	−20	74	95 : 5	R	91
30b	(naphthyl)	H	−20	98	95 : 5	R	92
30c	(naphthyl)	H	0	97	93 : 7	R	88
30d	t-Bu	H	−20	96	96 : 4	R	99.4
31a	t-Bu	H	−20	98	95 : 5	S	99.4

the shielding of one face over the other may be dependent more on mere steric crowding than on π-π-aryl/acrylate-orbital overlap. This was sustained by the results obtained with the acrylates from neopentyl ethers **27e**, **30d**, and **31a**. These proved to be far superior for chiral inductions (97–99.4% de) than any of the previously examined dienophiles. Thus, the alcohols **30** and **31** constituted "the first reported chiral auxiliaries which on acrylate/ cyclopentadiene-cycloaddition afford efficiently and predictably either (2R)- or (2S)-adducts, respectively, with virtually quantitative chiral induction."[22] The reaction of **32** with 1,3-butadiene in the presence of TiCl$_4$ produced the (R)-cyclohexenyl ester **33** (98% yield, >95.6% de).[23]

32   33

The results from **32** have been rationalized by a staggered conformation of the neopentyl side chain, which results in the steric shielding of the *re*-face of acrylate. In comparing the results of the acrylates derived from the chiral auxiliaries **27e** and **30d**, the increased efficiency was attributed to the buttressing effect of the C-10 methyl, which causes the ether side chain to be forced closer to the acrylate.[22] The arrangement of the acrylate in an *s-trans* orientation in the Lewis acid-promoted cycloadditions has been supported by recent calculations by Houk and co-workers.[24]

32

Oppolzer and Chapuis used the remarkable *si*-face selectivity exhibited by auxiliary **30d** in an enantioselective synthesis of (−)-β-santalene.[25] Bertrand had earlier described a synthesis of racemic β-santalene employing the Diels–Alder reaction of an allenic ester and cyclopentadiene to assemble the basic skeleton. Working along the same lines, Oppolzer planned to utilize allenic ester **34** to establish the required absolute configuration. Alcohol **30d** was first acylated with bromoacetyl bromide. Transformation into the phosphonium salt by reaction with triphenylphosphine and Wittig-type reaction of this salt with acetyl chloride yielded allenic ester **34** in 53% overall yield. Diels–Alder reaction with cyclopentadiene in the presence of $TiCl_2(Oi\text{-}Pr)_2$ at − 20°C provided adduct **35** in 98% yield (99% de; *endo:exo* = 98:2). The cycloadduct was converted by a series of steps to optically pure (−)-β-santalene (**36**). The chiral auxiliary was recovered in 95% yield following the LiAlH₄ reduction.

An investigation of the acrylates prepared from alcohols **37** and **38**, with the oxygen atom of the neopentyl ether of auxiliaries such as **27e** replaced by a methylene group, in the $TiCl_2(Oi\text{-}Pr)_2$-catalyzed reactions with cyclopentadiene was conducted. The acrylate from **37** produced a 75% yield of the adduct, which was 89% *endo* and had the (*S*)-configuration at C-5 with a 66% de, whereas the one from **38** gave an 89% yield of the (*R*)-configured adduct with a 94% de and was 92% *endo*.[26]

In a quest for alternative auxiliaries that would be more readily accessible, impart crystallinity to the dienophiles and the adducts, and be easily and efficiently regenerated, Oppolzer examined a number of derivatives of the readily available (+)- or (−)-camphor-10-sulfonic acids. The first two in this series to be examined were the sulfonamides **39**[27] and **40**.[28] They were efficiently prepared in two steps from (+)-camphor-10-sulfonyl chloride by amidation with diisopropyl- or dicyclohexylamine and subsequent L-selectride reduction. These were then converted to the acrylates **41** and **42** by the esterification conditions developed by Mukaiyama. When these dienophiles were subjected to $TiCl_2(Oi\text{-}Pr)$-mediated Diels–Alder reaction with cyclopentadiene, they afforded the cycloadducts **43** (98% yield, 97% *endo*, 88% de) and **44** (97% yield, 96% *endo*, 93% de). The crystalline adducts could easily be purified in excellent overall yield to 100% endo and 99% de by one or two recrystallizations. The auxiliaries were conveniently recovered following LiAlH$_4$ reduction of the adducts by simple crystallization. In the case of **44**, the sulfonamide **40** was recovered in 94% yield and (+)-**10** was isolated in 93% yield.[28] The uncatalyzed reaction of **42** with cyclopentadiene produces a 2.9:1 mixture of endo (35% de) and exo (13% de) adducts, respectively.[29]

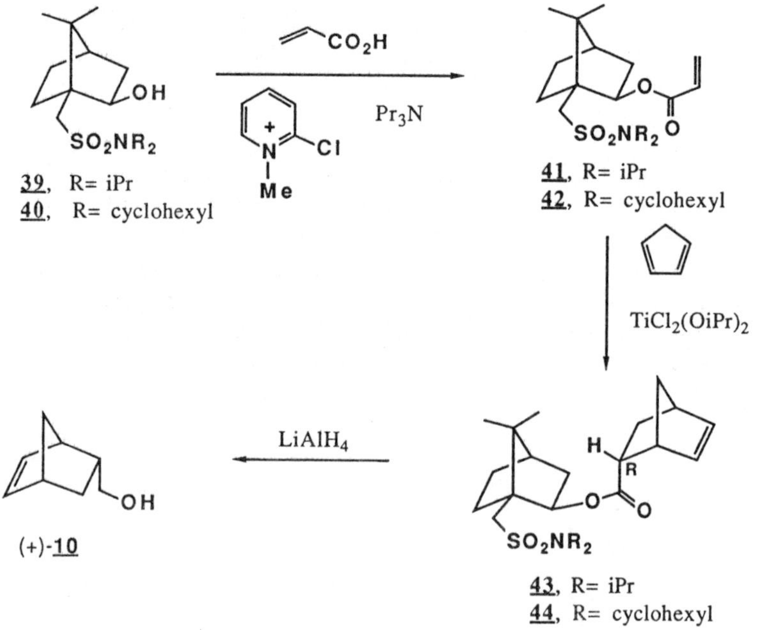

39, R= iPr
40, R= cyclohexyl

41, R= iPr
42, R= cyclohexyl

TiCl$_2$(OiPr)$_2$

(+)-**10**

43, R= iPr
44, R= cyclohexyl

These results were readily interpreted after an examination of the X-ray crystal structure of **42** revealed an antiplanar arrangement of the acrylate carbon–carbon double bond and the carbonyl. With the lone pair of electrons on the planar nitrogen atom bisecting the O—S—O angle, one of the cyclohexane rings is forced directly over the C$_\alpha$-*re*-face of the olefin.

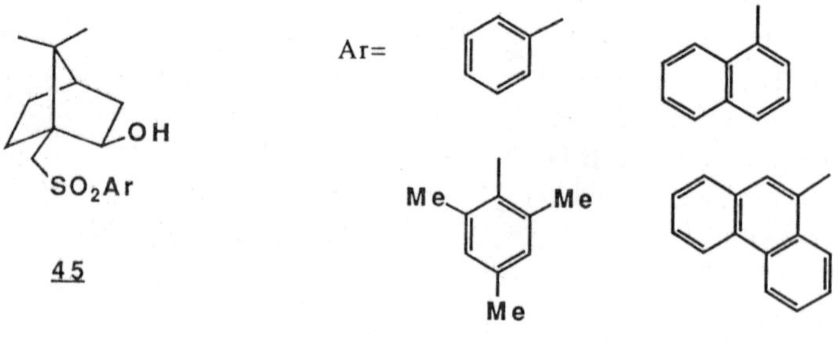

Ar=

45

The acrylates of sulfones **45**, derived from (+)-camphor-10-sulphonic acid, exhibit only moderate diastereoselectivities (64–69%) in their cycloadditions. This inferior stereodifferentiation will most likely preclude their general use in the asymmetric Diels–Alder reaction.[27,30]

A second generation of chiral auxiliaries derived from the camphor-10-sulfonic acids are the sultams **47** and **48**, which can be prepared from (+)- and (−)-camphor-10-sulfonic acid, respectively.[31] The sequence of transformations necessary for this conversion is illustrated for **47**. The route begins with the amidation of sulfonyl chloride **46** with NH_3. The amide is subjected to base-catalyzed cyclization to afford the intermediate imine, which is reduced to the sultam with $LiAlH_4$. The sultams are conveniently acylated by sequential treatment with NaH and then the appropriate acyl chloride to provide the highly crystalline *N*-acryloyl sultam **49** and the *N*-crotonoyl sultam **50**.

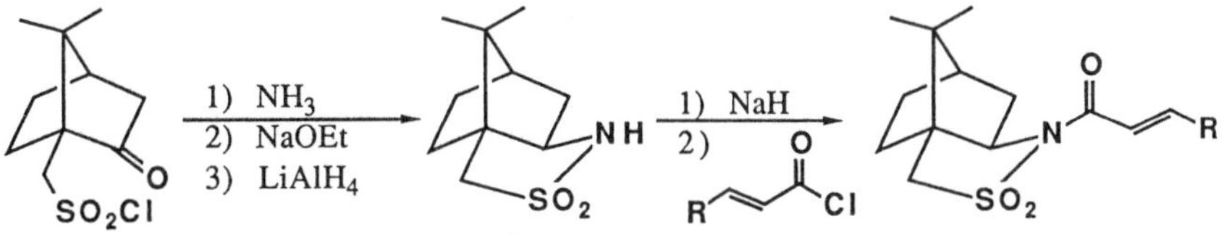

(+)-**46**

1) NH_3
2) NaOEt
3) $LiAlH_4$

47

1) NaH
2)

49, R= H
50, R= CH₃

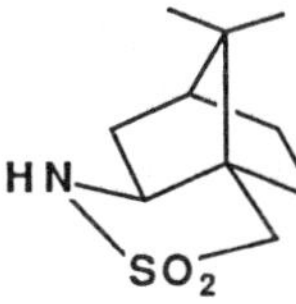

48 prepared from (-)-**46**

The X-ray crystal structure of **50** is shown below. Close scrutiny of the structure revealed a number of disturbing structural features: (1) the carbon–carbon double bond and the carbonyl group were synplanar, (2) the carbonyl and the SO_2 group were in an anti arrangement, and (3) the nitrogen atom was slightly pyramidal. All these factors seemed to indicate there would not be strong dienophile activation, and, even worse, the π-face differentiation in the Diels–Alder reaction would not be good.

However, when **49** and **50** were submitted to Lewis acid-mediated reaction with cyclopentadiene, a facile cycloaddition ensued at −130 and −78°C, respectively. The cycloadducts **51** and **52** were produced with excellent endoselectivity and diastereoselectivity. Even the less reactive 1,3-butadiene underwent facile cycloaddition at −78°C. An examination of the efficiency of various Lewis acids revealed $TiCl_4$ and $EtAlCl_2$ to be the most effective in terms of reaction rate, endo-selectivity, and chiral efficiency. It was also demonstrated that 1.5 mol equivalents Lewis acid should be employed rather than 0.5 mol equivalent (Table 6).

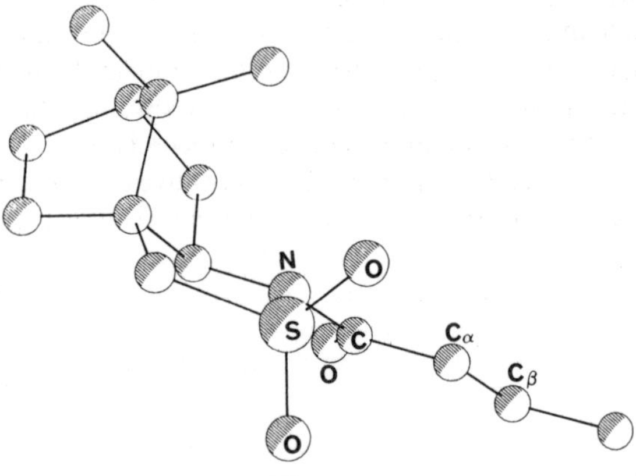

50.

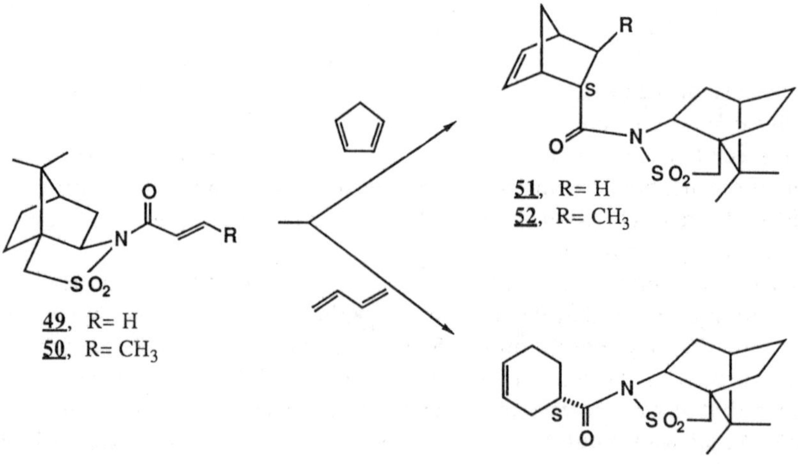

51, R= H
52, R= CH$_3$

49, R= H
50, R= CH$_3$

53

Table 6. Asymmetric Diels–Alder Reactions of **49** and **50**

Dienophile	Equivalent Lewis acid	Temperature (°C)	Adduct	% Yield	% endo	% de
49	0.5 TiCl$_4$	−130	51	87	96.3	91
49	1.5 TiCl$_4$	−130	51	89	97	94
49	1.5 BF$_3$·Et$_2$O	−130	51	58	89	51
49	0.5 EtAlCl$_2$	−130	51	85	94	85
49	1.5 EtAlCl$_2$	−130	51	96	99.5	95
50	0.5 TiCl$_4$	−78	52	98	99	93
50	1.5 EtAlCl$_2$	−78	52	91	96	98
49	1.5 EtAlCl$_2$	−78	51	93	—	97

The remarkable acceleration and diastereoselectivity of the Lewis acid-mediated cycloadditions of these sultams have been explained in terms of a chelated reactive intermediate. Chelation of the metal between the carbonyl and the SO_2 group locks the acylated sultam into the more favored conformation **53**. In this conformation the diene is forced to approach the less sterically congested C_α-re-face. The alternative conformation **54** is not as favorable because of the steric repulsion between the β-carbon of the dienophile and C-3.

Because of the highly crystalline nature of the adducts, the diastereomeric purity can be conveniently raised to 99% by simple recrystallization. The chiral auxiliary can be nondestructively removed by either $LiAlH_4$ reduction, if the alcohol is required, or saponification with LiOH, if the carboxylic acid is desired. The hydrolysis to the acid proceeds *without epimerization*. The auxiliary can be recovered in ~90% yield.[31]

Crotonoyl sultam **55**, derived from (−)-camphor-10-sulfonic acid, has been utilized by Vandewalle in the total synthesis of (−)-1-O-methyl loganin aglucone (**57**). Cycloaddition of **55** with cyclopentadiene in the presence of $TiCl_4$ produced the adduct, which was processed into alcohol **56**. It was shown that alcohol **56** was >97% enantiomerically pure. Using the methodology Vandewalle previously developed for the racemic synthesis, **56** was converted to the final target.[32]

In 1983, Masamune and co-workers described the design and use of the chiral dienophiles **58** and **59**.[33] These are prepared from the resolved carboxylic acids by addition of the appropriate vinyllithium reagent. These dienophiles react readily with cyclopentadiene at −20°C *in the absence of a*

Lewis acid catalyst to produce the cycloadducts **60** and **61** with excellent diastereofacial selectivity. In the case of **59**, the selectivity of >100:1 was the highest ratio obtained for an *uncatalyzed* reaction. The selectivities have been explained by invoking a five-membered chelate, which results from the intramolecular hydrogen bonding between the α-hydroxyl and the carbonyl. This restricts the free rotation and forces the carbonyl and the olefin of the enone into a synplanar conformation. In this conformation, the diastereo-topic faces are readily distinguishable and diene is thus directed to the face opposite the cyclohexyl- or *t*-butyl group. The silylated versions of **58** and **59** do not display good diastereofacial selection in their Diels–Alder reactions with cyclopentadiene.

58, R= cyclohexyl
59, R= t-Bu

60, R= cyclohexyl
61, R= t-Bu

Following the uncatalyzed studies, Masamune and co-workers examined the effect of Lewis acid catalysis on the cycloadditions of **59** with various dienes.[34] Not surprisingly, in the presence of Lewis acids the reaction times are shortened and the endo:exo ratio for the adducts improves. The results indicated the Lewis acids $Ti(Oi\text{-}Pr)_4$, $ZnCl_2$, and $BF_3 \cdot Et_2O$ were effective in catalyzing the cycloadditions, with the later two being the Lewis acids of choice. Again, these results have been explained by invoking the coordi-nation of the Lewis acid with the α-hydroxyketone to produce a rigid five-membered ring chelate.

The dienophiles **59** and **62** have been applied to the enantioselective syntheses of a number of natural products previously synthesized in racemic form. The reaction of (*S*)-**59** with 1,3-butadiene in the presence of $ZnCl_2$ produced adduct **63** (83% yield, 98% de). This was converted to alcohol **20**, whose enantiomer was previously converted to *R*-(−)-sarkomycin (**22**). The transformation of **63** to **20** involved DIBAL reduction, $NaIO_4$ cleavage of the diol, and DIBAL reduction of the resulting aldehyde. This sequence was accomplished in 58% overall yield. The cycloaddition of (*S*)-**59** with 1,4-diacetoxybutadiene in the presence of $BF_3 \cdot Et_2O$ furnished **64** in 72% yield as the exclusive stereoisomer. This has been converted in a series of six steps to (−)-shikimic acid (**65**). In a synthesis of (+)-pumiliotoxin (**69**), (*R*)-**62** was reacted with butadiene carbamate **66** in the presence of $BF_3 \cdot Et_2O$ to produce **67** in 95% yield (98% de). Reduction of **67** with $LiBH_4$ and $NaIO_4$

cleavage of the diol yielded aldehyde **68**, which was converted by published procedures to **69**.[34]

The dienophile (*S*)-**59** was reacted with the chiral dienes (*R*)-**70** and (*S*)-**70** in the presence of BF$_3$·Et$_2$O to examine the possibility of double asymmetric induction in the Diels–Alder reaction. The reaction with (*S*)-**70** produced a 73% yield of adduct **71** with >130:1 stereoselection. With (*R*)-**70**, the adduct **72** was obtained in 70% yield with 35:1 stereoselection. Both of the adducts have identical absolute configurations at C-1 and C-2, which is the direct result of the chirality of **59**. Thus, the stereochemistry is controlled by the choice of either (*R*)- or (*S*)-**59**.

The difference in the stereoselections (130:1 vs. 35:1) for the two adducts corresponds to the matched and mismatched pairs, respectively, in the cycloaddtion.[34] Although the dienophiles designed by Masamune are capable of creating new chiral centers in a predictable manner, the need to resolve the starting acids and the ultimate destruction of the chiral auxiliary will undoubtedly preclude the general applicability of this protocol.

73, R$_2$= iPr
74, R$_2$= CH$_2$Ph

Evans has reported the preparations and cycloadditions of the chiral α,β-unsaturated carboximides, **73**, **74**, and **75**, derived from (S)-valinol, (S)-phenylalanol, and (1S, 2R)-norephedrine, respectively.[35] Optimal results are obtained when the reactions are conducted in the presence of 1.4 equivalents of Et_2AlCl. Under these conditions, the dienophiles underwent efficient cycloaddition with cyclopentadiene in 2 min at $-100°C$. The cycloadducts were obtained with high endo selectivity ($>48:1$), high diastereomeric purity ($>93\%$), and in excellent yield ($>78\%$) (Table 7). It is interesting to note the reversed diastereoselectivity exhibited by the dienophiles derived from norephedrine.

Table 7. Et_2AlCl Catalyzed Reactions of **73–75** with Cyclopentadiene

Dienophile		Adduct	% endo	Isolated % yield	% de (purified)
73a	$R_1 = H$	**76**	>99	81	>98
73b	$R_1 = CH_3$	**76**	98	82	>98
74a	$R_1 = H$	**76**	>99	78	94
74b	$R_1 = CH_3$	**76**	98	83	98
75a	$R_1 = H$	**77**	>99	82	>98
75b	$R_1 = CH_3$	**77**	98	88	>98

The results have been attributed to the complexed ion pair **78**, which locks the carboximide into a conformation with the olefin and the carbonyl in a synplanar arrangement. In this conformation, the stereochemical outcome of the reaction is explained by assuming the C-4 substituent directs the diene to the opposite face of the dienophile. The electron-deficient nature of the complex can be used to rationalize the increased reactivity of these dienophiles as well.

In fact, less reactive acyclic dienes also add readily to these dienophiles. For the acyclic dienes, the dienophiles **74**, derived from phenylalanol, exhibit the best diastereoselectivities (Table 8). In all of these cases, the chiral auxiliaries can be efficiently and nondestructively regenerated by transesterification with lithium benzyloxide. Adduct **80**, prepared by the cycloaddition of isoprene and **74a**, was utilized in a synthesis of (R)-$(+)$-α-terpineol (**81**).

This was transesterified with LiOBn, and the resultant ester was reacted with methylmagnesium bromide to give **81**. This also verified the sense of asymmetric induction in the cycloaddition.

Table 8. Reactions of **74** with Isoprene and Piperylene Catalyzed by Et$_2$AlCl

Dienophile	Diene	Temperature (°C)	Isolated % yield	% de (purified)
74a, R$_1$ = H	Isoprene	−100	85	>98
74b, R$_1$ = CH$_3$	Isoprene	−30	83	>98
74a, R$_1$ = H	Piperylene	−100	84	>98
74b, R$_1$ = CH$_3$	Piperylene	−30	77	>98

The exceptional selectivity induced by the auxiliary derived from phenylalanol was believed to be the result of π-stacking with the phenyl ring possibly participating in a charge-transfer interaction.

In an elegant study of phenyl-substituted phenylalanol auxiliaries Evans *et al.* have shown there was little evidence to support a charge-transfer interaction. The conclusion was reached that the "enhanced steric effect" was caused by dipole–dipole and van der Waals attractions rather than charge-transfer.[35c]

Table 9. Diels–Alder Reactions of **83** with Cyclopentadiene

Solvent	Temperature (°C)	Equivalent Lewis acid	**84a : 84b**	**84c : 84d**	(**84c** + **84d**): (**84a** + **84b**)
n-Hexane	0	—	85:15	80:20	1.68:1
CH_2Cl_2	0	—	68:32	58:42	3.22:1
CH_2Cl_2	−45	1.1 $TiCl_4$	—	15:85	16:1
CH_2Cl_2	−45	1.1 $BF_3 \cdot Et_2O$	—	66:34	12:1
CH_2Cl_2	−63.5	1.1 $EtAlCl_2$	—	67:33	16:1
CH_2Cl_2	−63.5	0.75 $TiCl_4$	—	7:93	82:1
CH_2Cl_2/n-hexane	−63.5	0.75 $TiCl_4$	—	7:93	39:1

On the basis of work done earlier with the carbamoyl acrylate derivative of isoborneol **82**, which seemed to be exerting an electronic, as opposed to a steric, face-differentiating effect in its uncatalyzed reactions,[36] an examination of the acrylate **83** derived from (S)-ethyl lactate was undertaken.[37a] The lactate ester was chosen because the methyl group and the ester substituent of the lactate possess approximately equivalent steric size but have distinctly different electronic character. This acrylate also had the additional advantage of having two donor centres for the formation of chelated complexes with Lewis acids. In the uncatalyzed reaction with cyclopentadiene, the diastereofacial selectivity was found to be highly solvent dependent (Table 9). The best results in terms of diastereoselectivity were obtained with n-hexane as the solvent (60% de), but under these conditions the endo:exo ratio was not very useful (1.68:1). However, in the presence of TiCl$_4$, this dienophile reacted with cyclopentadiene to provide the cycloadduct **84d** with a much improved diastereoselectivity (86% de) and endo:exo ratio (39:1). Helmchen and co-workers have utilized adduct **84d** in the syntheses of the methyl esters of (R)- and (S)-sarkomycin (**22**).[37b] The diastereofacial selectivity can be reversed to that seen in the uncatalyzed reactions by switching to EtAlCl$_2$ or BF$_3$·Et$_2$O as the Lewis acid, although the level of induction is not as high (~33% de). The auxiliary is conveniently removed by saponification with LiOH without any epimerization.

The stereodichotomy exhibited by this dienophile, which depended on the Lewis acid employed, could not be explained by any of the known concepts concerning the asymmetric Diels–Alder reaction. This led to an examination into the nature of the complex formed between the dienophile and the Lewis acid. The work culminated in the isolation of crystals of the 1:1 complex **83a** of acrylate **83** with TiCl$_4$. The X-ray crystal structure of this complex revealed a number of interesting features. The olefin and the carbonyl of the acrylate are synplanar, not antiplanar as expected. Also, the diastereoselectivity could now be explained by the shielding of the C_α-*re*-face of the acrylate by one of the *chlorine atoms of the Lewis acid*. The dichtomy displayed the different Lewis acids can now be explained by assuming tetracoordinating Lewis acids such as EtAlCl$_2$ and BF$_3$ result in the formation of complexes 85, which would shield the C_α-*si*-face of the dienophile.[38]

The information obtained from the crystal structure of the Lewis acid complex 83a led to the design of an auxiliary that might help stabilize the intermediate complex. Examination of the structure revealed a torsional angle of 20° for the lactate backbone. Extrapolation of these data led to the postulation that a cyclic structure such as 86 should have a coordination geometry similar to that of 83a. The cyclic structure should have the added advantage that the entropy balance between competing monodentate complexes should be more favorable for the cyclic compound. With the aforementioned thought in mind, the readily available D-pantolactone (87) was studied as a chiral auxiliary. The acrylate 88 in combination with as little as .1 equivalents of TiCl$_4$ reacted with cyclopentadiene, 1,3-butadiene, and isoprene affording the adducts **89–91** in excellent yield with diastereoselectivities of >86%. The crystallinity of the adducts allowed for simple purification by recrystallization. After a few crystallizations the diastereoselectivities could be increased to >99%. The auxiliary can be nondestructively removed by saponification with LiOH without epimerization of the cycloadduct.[39]

83a   **86**   **87**

88   TiCl₄   **89**   LiOH

90, R= H
91, R= CH₃

Both enantiomers of Matsutake alcohol **93** have been prepared using adduct **92** which is obtained in 82% yield from the reaction of **88** with anthracene.[37]

88   **92**   (-)-(R)-**93**

Two arabinose-based *si*-face directing chiral auxiliaries have recently been developed. The derived acrylates **94** and **95** were reacted with the diene shown in Table 10 in the presence of EtAlCl₂ to afford moderate yields of the cycloadducts possessing the (R)-configuration with diastereomeric excesses ranging from 10 to 70%.[40]

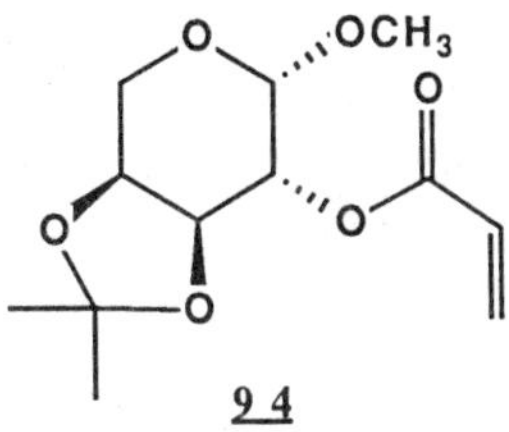

9 4 **9 5**

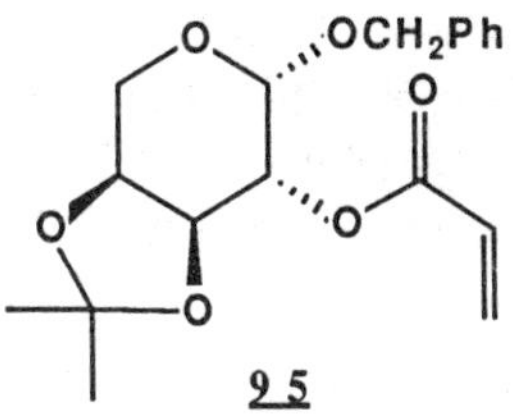

Table 10. Et$_2$AlCl Catalyzed Reactions of **94** and **95**

Diene	Dienophile	Diastereoselectivity	% Yield
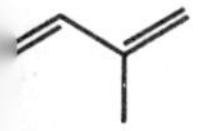	94	81 : 19	56
	95	64 : 36	40
	94	73 : 27	68
	95	64 : 36	58
	94	79 : 21	65
	95	68 : 32	50
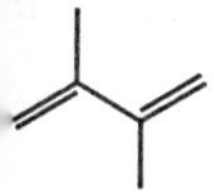	94	66 : 34	51
	95	55 : 45	39
	94	85 : 15	48
	95	75 : 25	37

The strategy of using carbohydrate derivatives as chiral templates in the Diels–Alder reaction was first investigated by Jones, using **96** as the dienophile.[41] Under thermal conditions with a number of acyclic dienes, the reaction reportedly produced mixtures of adducts epimeric at the C-1 hemiacetal center. Cyclic dienes selectively provided adducts **98**.

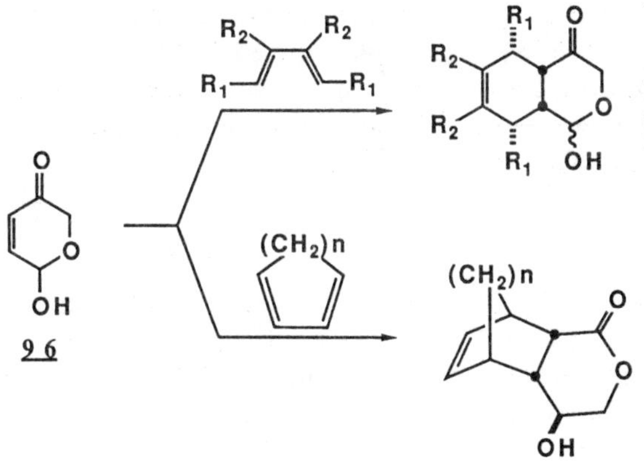

97a, R$_1$ = R$_2$ = H
97b, R$_1$ = H; R$_2$ = CH$_3$
97c, R$_1$ = CH$_3$; R$_2$ = H

9 6

98a, n = 1
98b, n = 2

Hutchinson and co-workers, in a synthetic approach to iridoids and alkaloids, employed the cycloaddition of 96 and its methylated derivative 9 with 1,3-butadiene. With **99**, only adduct **100** was reportedly isolated.[42] This structural assignment was later altered to *epi*-**100**.[42b]

99   100 °C   **100**   epi-**100**

The cycloadditions of **99** have been reexamined under thermal and high pressure conditions by Jurczak. The utilization of high pressures allows for the employment of milder reaction conditions, which usually results in an increase in the diastereoselectivity. Under strictly thermal conditions, 9 reacts with 2,3-dimethyl-1,3-butadiene to deliver **101** as the major product and **102** as the minor component in a ratio of 92:8. When the reactions were examined under the high-pressure conditions, **101** appeared to be the exclusive product.[43]

99   **101**   **102**

In a preparation of annelated pyranosides, Fraser-Reid and co-worker have disclosed a Lewis acid-mediated Diels–Alder reaction with a carbo hydrate derivative. The reaction of **103** with 1,3-butadiene in the presenc of excess $AlCl_3$ produced **104** as a single adduct in 80–90% yield. The Lewi acid-catalyzed conditions have the advantage over the high-pressure con ditions in that the reactions can be performed on a preparative scale withou the use of a specialized apparatus.[44]

103 $AlCl_3$ **104**

Rahman and Fraser-Reid have used the Diels–Alder strategy in a total synthesis of (+)-actinobolin (**107**).[45] The basic ring system was assembled via the cycloaddition reaction between enone **105** and Danishefsky's diene under thermal conditions.[46] This resulted in a 93% yield of cycloadduct **106** as the sole product. This was converted in a number of steps to (+)-actinobolin (**103**).

105   **106**   **107**

The cycloaddition of levoglucosenone (**108**) with 1-acetoxy-1,3-butadiene has been employed by Isobe *et al.* in a synthetic approach to tetrodotoxin (**110**). The cycloadduct has been converted into the key intermediate **109**, which will ultimately be converted into the final target.[47]

108   **109**

110   **111**

The chiral butenolide **112a**, which was prepared from D-(+)-ribono-lactone, reacted with 1,3-butadiene in the presence of AlCl$_3$ to render the adduct **113a** in 75% yield as the exclusive product.[48] The stereochemical integrity of **113a** was confirmed by ^{1}H-NMR studies in the presence of chiral shift reagents. In an independent study, the cycloadditions of a number of similar butenolides derived from D-ribonolactone with 1,3-butadiene under thermal conditions produced the isomerically pure adducts **113b–g**. The reactions of cyclopentadiene with lactones **112b,e,f** at 80–120°C gave endo: exo ratios of about 75:25 in yields of 55–85%.[49]

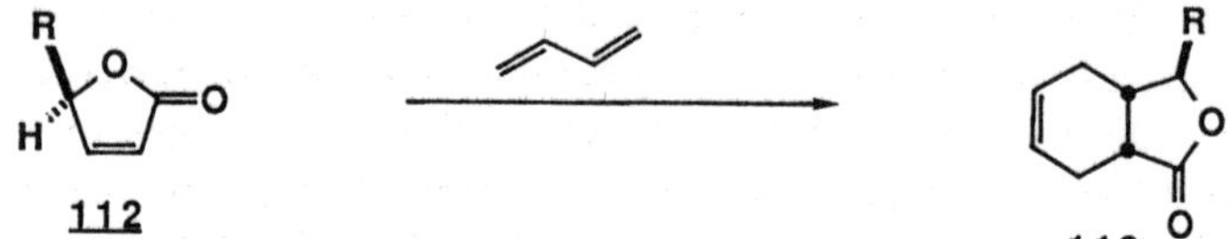

112

113

a, R= Ph₃COCH₂-
b, R= CH₃
c, R= CH₂OH
d, R= CH₂OAc

e, R= CH₂OCH₃
f, R= CH₂OBn
g, R= CH₂SPh

Takano *et al.* have employed the endo-cycloadduct **114** from the reaction of **112c** and cyclopentadiene in an enantiodivergent route to the two enantiomers of β-santalene (**36**) and epi-β-santalene.[49d]

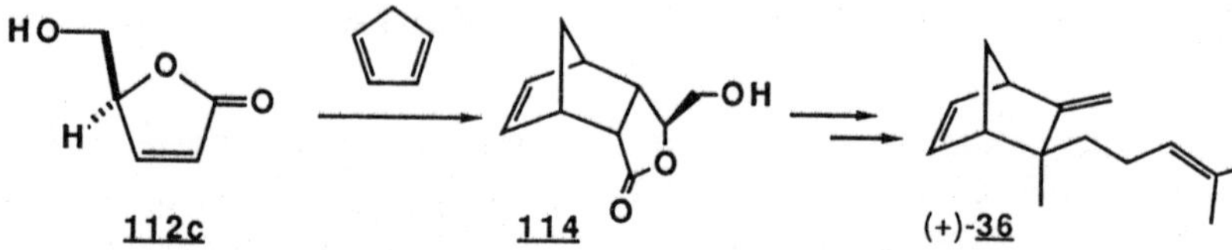

112c   **114**   (+)-**36**

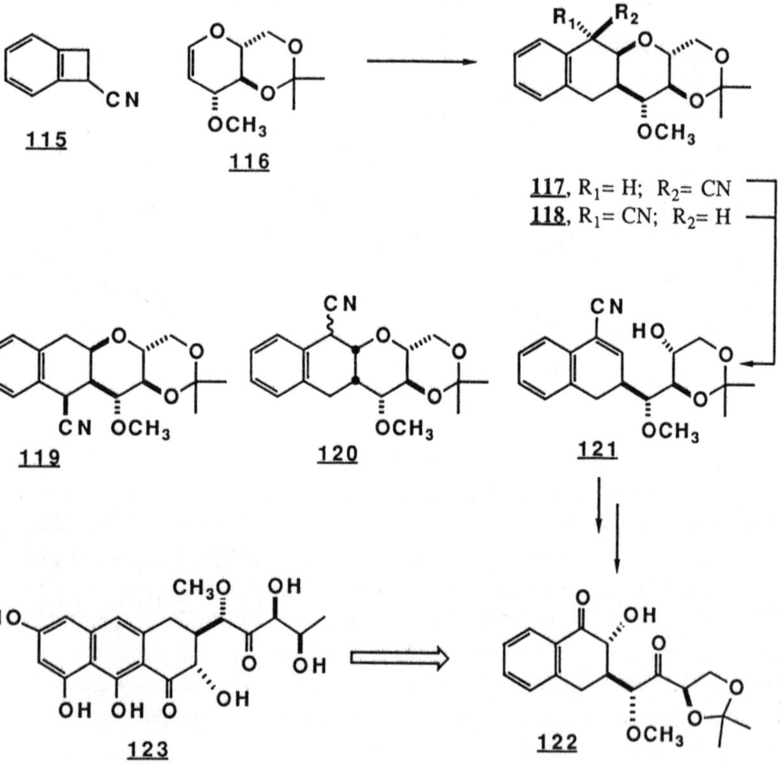

115   **116**

117, R₁= H; R₂= CN
118, R₁= CN; R₂= H

119   **120**   **121**

123   **122**

A model study reported by Franck and John, directed at the preparation of the aglycone of the aureolic acid class of antitumor antibiotics [i.e., olivin (123)], utilized the Diels–Alder reaction of an *o*-quinone methide and a glucal derivative. The reaction of cyanocyclobutane **115** and the glucose derivative **116** afforded the cycloadducts **117** (20%), **118** (45%), **119** (9%), **120** (14%), and **121** (4%). The regioselectivity was 9:1 in favor of the desired regioisomer with a stereoselectivity of 5:1 within this series. The adducts **117** and **118** can be ring opened to yield **121**. This was converted in a sequence of reactions to **122**, which contains the crucial stereochemical features of the aureolic acids.[50]

Trapping of an *o*-quinone methide with a carbohydrate-based dienophile has been extended to the acyclic unsaturated sugar derivatives **124–126**. These dienophiles afforded the opportunity to examine which of the two allylic substituents would control the diastereofacial selectivity in the cyclo-

124, R= OEt
125, R= CH$_3$

126

Steric Control / Orbital Control

127

128

129

128a, R= OEt; R$_1$= Ac; R$_2$=

128b, R= CH$_3$; R$_1$= Ac; R$_2$= **128a**

128c, R= CH$_3$; R$_1$= CH$_3$; R$_2$=

129a, R; R$_1$; R$_2$= **128a**
129b, R; R$_1$; R$_2$= **128b**
129c, R; R$_1$; R$_2$= **128c**

addition. On the basis of steric arguments, the diene should be directed to the *re*-face, away from the bulky sugar group and should result in the (*R*)-configuration at C-3 of the adduct. Alternatively, to minimize the secondary orbital antibonding effects, the allylic alkoxy group could direct the diene to the *si*-face and result in the (*S*)-configuration at C-3. Condensation of benzocyclobutene **127** with **124–126** produced the adducts **128** and **129** with selectivities of 1.2–4:1 in favor of **128**. The C-3 stereochemistry was proven to have the (*S*)-configuration via an X-ray structure of a lactone derived from adduct **128a**. These results were consistent with the orbital control hypothesis in which the LUMO–HOMO interaction of the syn alkoxy group was more favorable than that of the syn alkyl group. This resulted in direction of the diene to the *si*-face.[51]

The exo orientation of the carboxyl group was also consistent with the results obtained earlier by Horton and co-workers with the acyclic unsaturated sugar derivative **130**, derived from L-arabinose. Cycloaddition of **130** with cyclopentadiene under thermal conditions reportedly produced the single, optically pure adduct **131** in 65% yield after recrystallization.[52] These results have been held suspect because of the lack of HPLC or GC data and the fact that the optical rotation of ester **132** was compared with a published value for enantiomerically impure **132** ($\sim$25% ee).[8a]

The original report has since been amended because it was discovered that the commercial material used to prepare **130** was actually D-arabinose.[52b] The reaction of the dienophile **133** derived from D-arabinose with cyclopentadiene under thermal conditions produced **134** in 40% yield. The structure of **134** was unambiguously assigned via X-ray crystallography.

The thermal reaction of **130** and cyclopentadiene afforded a mixture of adducts **135**, **136**, **131**, and **137** in 33, 5, 18, and 12% yields, respectively. If this reaction was catalyzed by $AlCl_3$, adducts **135**, **136**, **131**, and **137** were obtained in 7, 15, 8, and 49% yields, respectively.

More recently, the uncatalyzed Diels–Alder reactions of dienophiles **138** and **139**, prepared from (R)-2,3-isopropylidene glyceraldehyde, with cyclopentadiene were reported. The condensation of **138** and cyclopentadiene yielded adducts **140** and **141** in a 92:8 ratio, respectively. The analogous reaction with **139** furnished a mixture of adducts **142** and **143** in a 60:40 ratio. The reactive conformation for the cycloaddition was assumed to be as depicted in **144**, as opposed to the alternatives **145** and **146**. In the case of **138** the endo directing effects of the carboxyl and the dioxolane are cooperative, whereas for **139** they roughly cancel each other out.[53]

A double-barreled version of this cycloaddition has been used in an enantioselective approach to loganin. The D-mannitol derived bis-unsaturated ester **147** was reacted with cyclopentadiene, under the influence of Et$_2$AlCl, to afford a 90% yield of a 3:2 mixture of adducts **148** and **149**.[54]

The absolute configurations of the adducts were determined by conversion to the known diol **150**. Adduct **148** was transformed into **151**, which was an intermediate in Vandewalle's synthesis of loganin (*vide supra*). The major product was believed to be the result of reaction via the anti-Felkin–Anh conformation **153**. The minor product was presumably formed through an alternative anti-Felkin–Anh conformer **154**. The Felkin–Anh conformation **152** suffers from nonbonded and dipole–dipole interactions.

In connection with projects aimed at the syntheses of enantiomerically pure norbornane-type terpenes, Magnusson and co-workers reported the reactions of the unsaturated aldehydes **155** and **156** with cyclopentadiene. These new chiral isoprene units were prepared in six steps from D- and L-arabinose, respectively. When **155** was reacted with cyclopentadiene in an uncatalyzed reaction at $-20°C$, it produced the adducts **157**, **158**, and **159** in a ratio of 82:18: <1, respectively. The addition of cyclopentadiene not unexpectedly occurred from the face opposite to the benzyloxy substituent. However, the major product was the result of an exo mode of addition.[55]

A photochemically mediated [4 + 2] cycloaddition of glycals and an azodicarboxylate has recently been utilized in a synthesis of 2-amino-2-deoxy carbohydrates. Thus, irradiation of the glycal **160** and dibenzylazo-dicarboxylate at 350 nm provided the adduct **161** as a single diastereomer. The diastereofacial selectivity in these cycloadditions appears to be controlled by the stereochemistry at C-3. The adduct was converted in a short sequence of reactions to the desired amino sugar **162**. In an analogous fashion, the glycals **163** and **164** were processed into amino sugars **165** and **166**.[56]

163   165

R= t-BuPh₂Si   R'= t-BuMe₂Si

164   166

The thermal Diels–Alder reaction of (−)-carvone (**167**) with butadiene was originally reported to give the two adducts **168** and **169** in a combined yield of 8%. The supposed major product **168** (6% yield) appeared to result from an addition *syn* to the isopropenyl group.[57] A reinvestigation of this reaction has resulted in a reversal of the assignment of these two products, with the structure of the major product now being ascribed to **169**, which results from an addition *anti* to the isopropenyl group. When the reaction was conducted in the presence of 0.1 equivalents of AlCl₃, a 40% yield of adduct **169** was obtained along with 3% of enone **170**. None of the minor component **168** from the thermal reaction was reported.[58]

A reinvestigation by Wenkert and co-workers of the AlCl₃ mediated cycloadditions of (−)-carvone with 1,3-butadiene rendered the following octalones **168–170** in 80–88% yield.[59] The major products again resulted from endo addition anti to the isopropenyl group. The yield of the cyclo-adducts from 1,3-butadiene was approximately twice that previously reported and included the product resulting from syn addition, also not previously observed. It is interesting to note the reaction with isoprene produced a regioisomerically pure mixture of stereoisomers **171** and **172**. This contrasts with the reported isolation of regioisomers from the uncatalyzed reaction.[60]

High-pressure technology was necessary to effect the [4 + 2] cycloaddition of the cyclopropyl-substituted cycloheptenone **173** with furan **174** in Smith's syntheses of (+)-jatropholones A (**176**) and B (**177**). All attempts with thermal or Lewis acid conditions met with failure. However, when the two components as a 1:1 neat mixture were subjected to 5 kbar of pressure, an 80% yield of adduct **175** as a single diastereomer was realized. Compound **175** was converted into (+)-**176** and (+)-**177** via a short sequence of transformations.[61]

The chiral maleic anhydride analogue **178** reacts with cyclopentadiene in an enantioselective thermal Diels–Alder reaction to produce a 65% yield of **179** (>97% endo, >96% de). Hydrolysis and Wittig methylenation provided the intermediate **180**, which has previously been used in the synthesis of racemic 6,7-dehydroaspedidospermidine systems. This dienophile has been reacted with other dienes to afford adducts with diastereomeric excesses generally greater than 96%. The only exception so far has been in the reaction with sulfolene, which produces a product of only 73% de.[62]

The reaction of 1-benzoyl-1-menthyloxycarbonylethene (181) with cyclopentadiene under thermal conditions resulted in no diastereomeric excess in any of the adducts. However, in the presence of a Lewis acid, preferably ZnCl$_2$, a >95% diastereomeric excess was achieved. The adducts 182 and 183 were isolated in 88% yield as a 4:96 mixture.[63]

Two menthyl auxiliaries were used in the cyclization of the (acetoxymethylene) malonate 184 with cyclopentadiene for the enantioselective preparation of carbocyclic analogues of C-nucleosides. In the presence of a catalytic amount of TiCl$_4$, 184 undergoes cyclization with cyclopentadiene to produce 185 as a mixture of endo:exo isomers (3:1) in 83% yield and >90% enantiomeric excess. The absolute configurations were established by conversion to the known lactone 186.[64]

The use of chiral sulfoxides for the transfer of chirality has been applied to the asymmetric Diels–Alder reaction. The first use of a chiral sulfoxide for this purpose was reported by Maignan and Raphael. The reaction of (+)-(R)-p-tolyl vinyl sulfoxide (187) with cyclopentadiene at 115°C produced a mixture of the four isomeric products 188 (8%), 189 (28%), 190 (42%), and

191 (22%), which were readily separable by chromatography. The adducts **189** and **190** were processed individually into the enantiomeric forms of dehydronorcamphor (**18**).[65] The formation of four diastereomers in the Diels–Alder reaction makes this process too inefficient to be of practical use.

In an effort to improve the diastereoselectivity in the cycloaddition process, Maignan *et al.* opted for the use of the more reactive methyl-(Z)-(+)-(R)$_s$-*p*-tolylsulfinyl-3-propenoate (**192**).[66] This dienophile was synthesized in a manner analogous to that employed by Koizumi for the preparation of a methylated analog of **192** (*vide infra*), with the Emmons–Horner reaction of the anion of (+)-(R)$_s$-dimethyl-*p*-tolylsulfinylmethane phosphonate and methyl glyoxylate. This furnished a separable mixture (1.86:1; *E:Z*) of (*E*)- and (*Z*)-isomers in 83% yield. When the (*Z*)-isomer was reacted with cyclopentadiene at 4°C, it afforded the two adducts **193** (93%) and **194** (7%). Unfortunately, no reactions of the major (*E*)-isomer were reported.

Prior to Maignan's work, Koizumi *et al.* delineated their study of the uncatalyzed asymmetric Diels–Alder cycloadditions of cyclopentadiene with *E* and *Z*-(R)$_s$-ethyl-2-*p*-tolylsulfinylmethylenepropionate (**195**). The Emmons–Horner reaction of ethyl pyruvate with the optically active sulfinyl phosphonate produced a 23% yield of *E*-(R)$_s$-**195** and a 10% yield of *Z*-(R)$_s$-**195**. The reaction of *E*-(R)$_s$-**195** with cyclopentadiene at 90°C gave adducts **196a**, **196b**, and **197a** in 63, 15, and 22% yields, respectively. None of the diastereomeric exo sulfoxide **197b** could be detected. The reaction of *Z*-(R)$_s$-**195** rendered the adducts **198a**, **198b**, and **199a** in 63, 2, and 35% yields, respectively. Again, none of the diastereomeric exo sulfoxide **199b** could be detected. The diastereoselectivity for the endo addition of *E*-(R)$_s$-**195** was

62% de, whereas the exo mode proceeded with 100% de. The endo and exo additions of Z-$(R)_s$-**195** resulted in a 94% de and 100% de, respectively. The absolute configurations of adducts **196a** and **199a** were determined by chemical correlation with the known esters (−)- and (+)-**200**.[67]

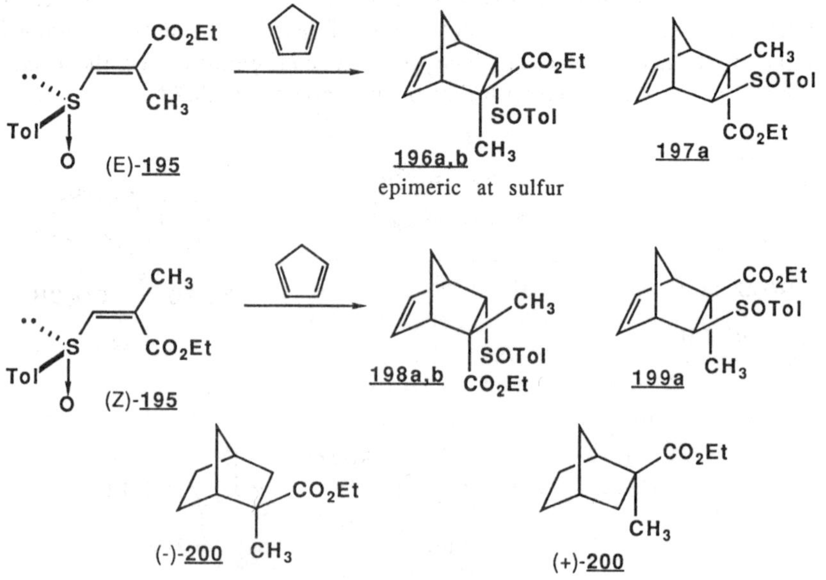

epimeric at sulfur

The results have been rationalized by an s-trans conformation of the S—O and olefinic bonds of the unsaturated sulfoxide as illustrated below for E-$(R)_s$-**195**. In this conformation, the diastereoselectivity has been explained in terms of the difference in steric bulk between the aromatic group and the lone pair of electrons on sulfur, with the preferred mode of addition for the diene being from the same side as the lone pair.

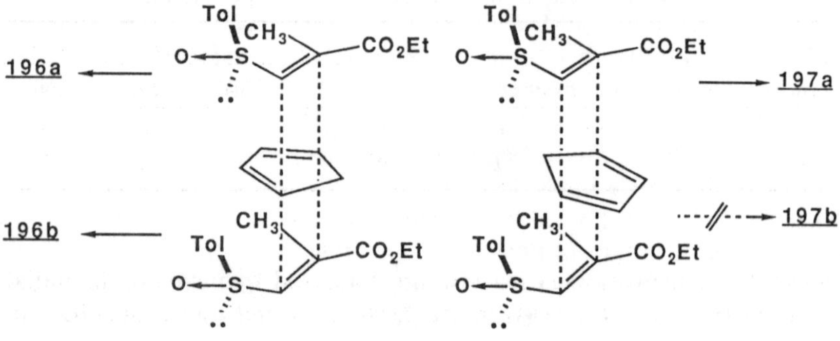

Koizumi's group has also reported the use of alkyl $(S)_s$-2-p-tolylsulfinyl-acrylates **201a** and **201b**. The reaction of **201a** with anthracene in the presence of $ZnCl_2$ afforded cycloadduct **202** as a single diastereomer. The cyclo-addition of **201b** with cyclopentadiene resulted in the formation of adducts **203–206**. Table 11 shows a dramatic change in the ratio of diastereomers when the reaction was catalyzed by $ZnCl_2$. The ratio was almost completely reversed in the Lewis acid-catalyzed reaction from that obtained in the noncatalyzed reaction. The absolute configuration for the cyclo-adduct **205** was correlated with that of the known acid **207**.[68]

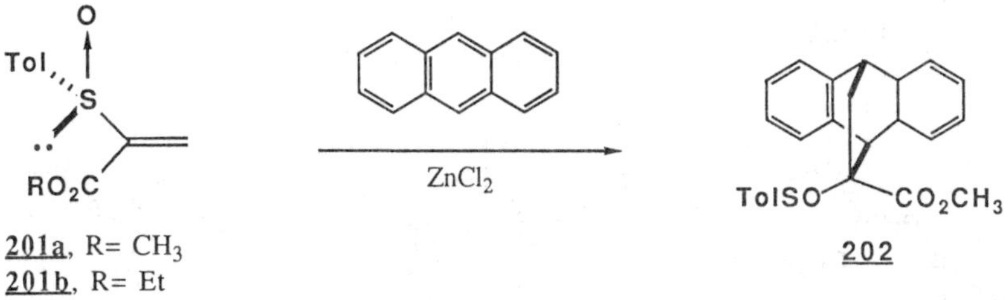

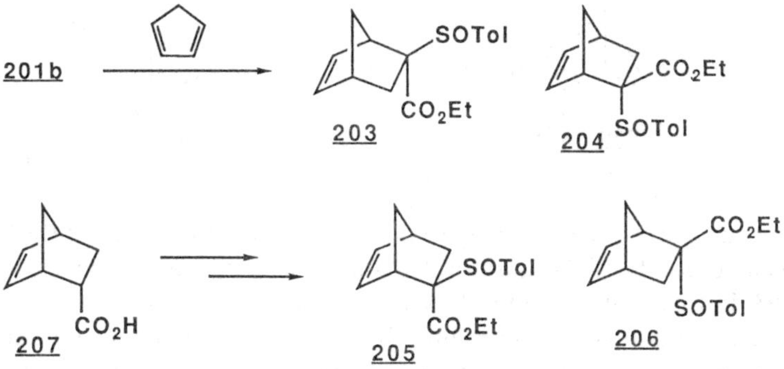

201a, R= CH₃
201b, R= Et

202

Table 11. Reactions of **201b** with Cyclopentadiene

		% Yield			
Lewis acid	*Temperature (°C)*	**203**	**204**	**205**	**206**
—	20	64	11	23	2
$ZnCl_2$	0	2	77	2	19

The results of the catalyzed reaction are explained by invoking the initial formation of a complex between the Lewis acid and the dienophile and subsequent attack of the diene from the least hindered side of the chelated structure **208**, as shown below.

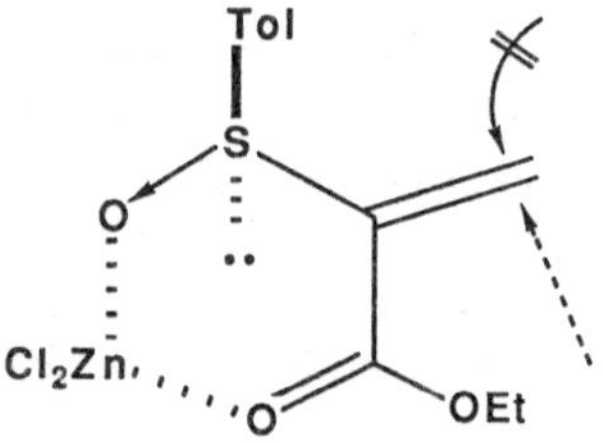

208

The preparation and use of (+)-(S,S)$_s$-1,1-bis(p-tolylsulfinyl)ethene (**209**) as a latent chiral ketene equivalent has recently been reported by Koizumi, and co-workers. Reaction of (+)-**209** with cyclopentadiene at 70°C provided an inseparable mixture of **210a** and **210b** in a ratio of 4:1. The absolute configuration was determined by conversion of the mixture into the known (+)-(1R)-norbornenone **18** with an enantiomeric excess of 54%. It was reported that addition of ZnCl$_2$ changed the diastereomeric ratio, but no details concerning this experiment were provided.[69]

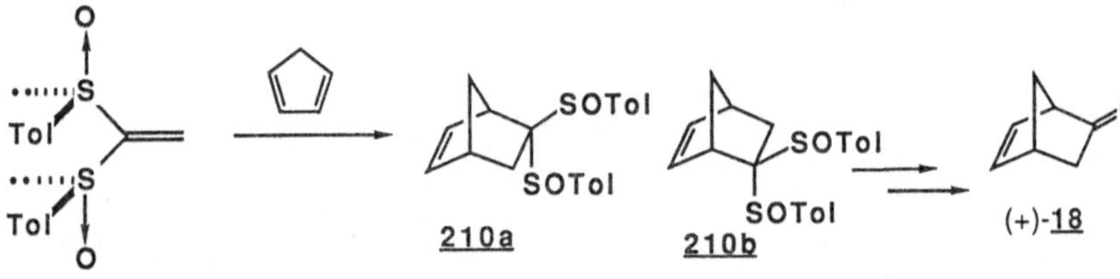

209   **210a**   **210b**   (+)-**18**

Maignan and Belkasmioui have recently developed a new chiral ketene equivalent. They used (R)-(+)-ethynyl-p-tolylsulfoxide (**211**) in an uncatalyzed reaction with cyclopentadiene to produce a 97% yield of a separable 70:30 mixture of adducts **212** and **213**, respectively. These were independently converted into the two enantiomers of norbornenone **18**.[70]

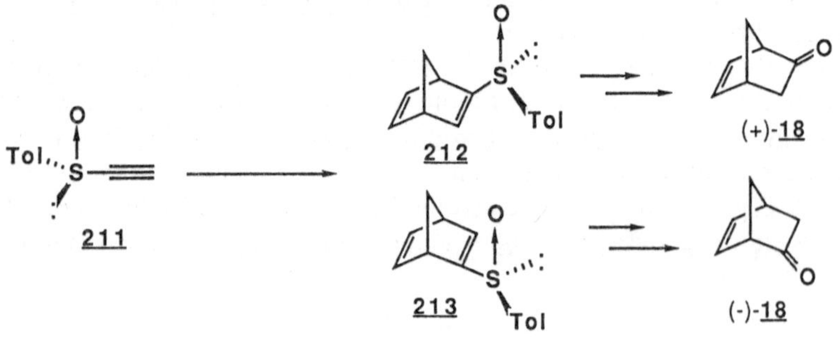

211   **212**   (+)-**18**

213   (-)-**18**

The use of $(S)_s$- and $(R)_s$-menthyl-3-(2-pyridylsulfinyl)acrylates **214** and **215** for the asymmetric preparation of 7-oxabicyclo[2.2.1]hept-5-ene-2-carboxylate derivatives via the Diels–Alder reaction has also been described by Koizumi and co-workers. These dienophiles are readily available from (+)-menthyl propiolate by conjugate addition of 2-mercaptopyridine, subsequent *m*-CPBA oxidation, and fractional crystallization of the diastereomeric sulfoxides. This gave **214** and **215** in 20 and 13% overall yield from the acetylenic ester. The Diels–Alder reaction of **214** and **215** with furan in the presence of Et$_2$AlCl afforded the adducts **216–219** in the yields shown. The absolute configurations of the major isomers were determined by correlation with the known alcohols (+)-**220** and (−)-**220**.[71]

An enantioselective total synthesis of (−)-glyoxylase I inhibitor (**223**) has been achieved with a slightly modified version of the pyridylsulfinyl acrylate **214**. It has been found that placement of a trifluoromethyl group at the 3-position of the pyridyl moiety greatly enhanced the reactivity of the derived sulfinyl acrylate. The preparation of $(S)_s$-**221** directly parallels that of the destrifluoromethyl derivative **214**. Following fractional crystallization of the 1:1 mixture of diastereomeric sulfoxides, $(S)_s$-**221** is isolated in 20% overall yield from menthyl propiolate. Condensation of 2-methoxyfuran and $(S)_s$-**221** at 0°C produced almost exclusively the endo adduct **222** (98% de). This was transformed by a series of reactions into (−)-glyoxylase I inhibitor (**223**).[72]

Cycloadditions of the vinyl sulfoxides derived from the hydroxythiol **224**, which can easily be prepared via the LiAlH$_4$ reduction of (+)-camphor-10-sulfonyl chloride, with cyclopentadiene have been reported by De Lucchi *et al.*[73] Michael addition of the mercaptan to the electron-poor acetylenes **225** produced the vinyl sulfides **226**. The Michael adducts from the sulfonyl acetylenes possessed (*Z*)-olefin geometry. In the case of the propiolic derivatives, either the (*E*)- or (*Z*)-isomer could be obtained by simply changing the base and the solvent in the Michael addition. The (*Z*)-isomer is prepared with Et$_3$N in methanol/water as solvent. The (*E*)-isomer can be obtained with 1,4-diazabicyclo[2.2.2]octane in acetonitrile or by irradiation of the (*Z*)-isomer at 350 nm. Highly diastereoselective hydroxyl-directed oxidation of the sulfides with *m*-CPBA in CH$_2$Cl$_2$ produces the sulfoxides **227** and **228**. Other oxidants proved to be less selective. Reaction of the vinyl sulfoxides with cyclopentadiene at 0°C afforded the adducts **229** and **230** (Table 12). A high degree of diastereoselectivity was observed in the cycloadditions of the (*Z*)-olefins. The cycloaddition of the corresponding (*E*)-isomers afforded in all cases mixtures of diastereomers. The absolute configurations of the adducts were determined by X-ray crystallography to be as shown.[74]

Table 12. Cycloadditions of **227** and **228** with Cyclopentadiene

X	Olefin stereochemistry	227 : 228	229 : 230
PhSO$_2$	Z	90 : 10	90 : 10
PhSO$_2$	Z	61 : 39	61 : 39
PhSO$_2$	E	90 : 10	Mixture
p-ClPhSO$_2$	Z	91 : 9	91 : 9
p-ClPhSO$_2$	Z	63 : 37	63 : 37
p-ClPhSO$_2$	E	100 : 0	Mixture
CO$_2$CH$_3$	Z	96 : 4	98 : trace
CO$_2$CH$_3$	Z	4 : 96	trace : 98
CO$_2$CH$_3$	E	100 : 0	Mixture

The rotomeric conformational preferences of the dienophiles **227b,c** and **228b,c**, determined by the analysis of NOE measurements, are shown below. In these conformations, the diene should be directed to the *re*-face of dienophiles **227b** and **227c** and to the *si*-face of dienophiles **228b** and **228c**. The X-ray analyses have confirmed that the diene does indeed add to these faces of the dienophiles. The diastereoselectivities have been explained in terms of a synergistic effect of the two chiral elements, the sulfoxide and the camphor auxiliary. The critical interaction was the hydrogen bonding between the sulfoxide and the alcohol, which imposed the necessary conformational rigidity into the system.

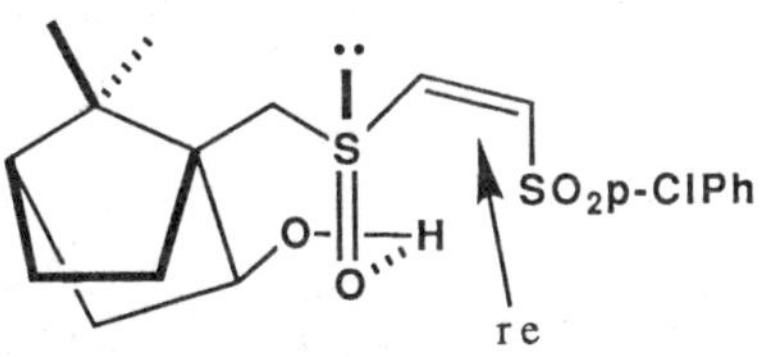

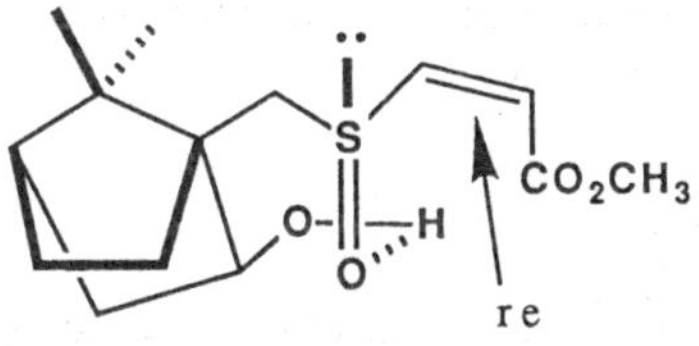

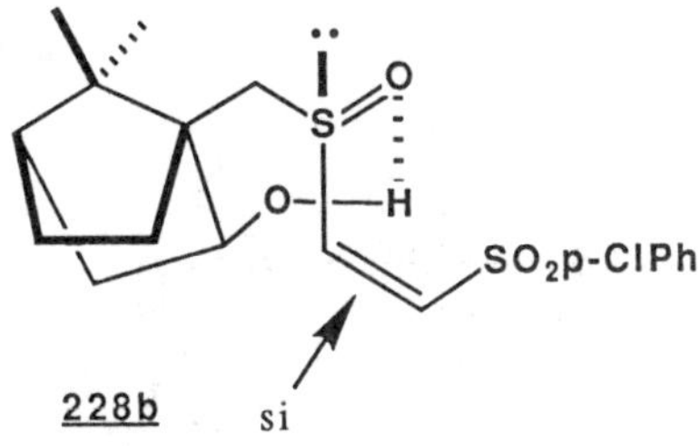

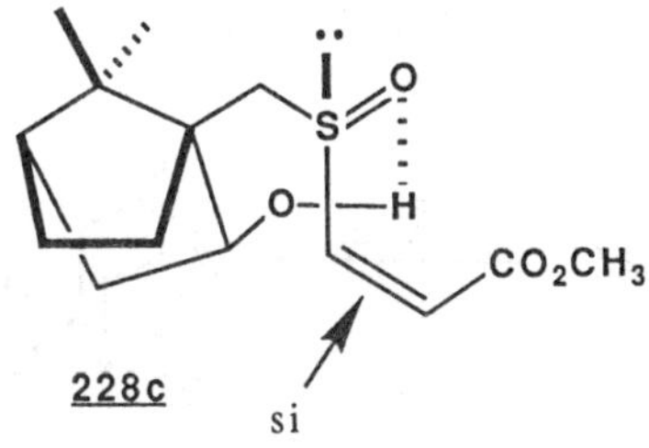

Kahn and Hehre recently proposed the diastereofacial selectivity in the cycloaddition reactions of achiral dienes and chiral vinyl sulfoxides is the result of the electrostatic preference for a nucleophilic diene to add to the face that would avoid the electron-rich lone pair on sulfur, even if it means adding to a face that is sterically encumbered. It is presumed the vinyl sulfoxide reacts from an s-cis conformation in which one face of the olefin is rendered electron rich by the lone pair on sulfur, whereas the other face is left relatively electron poor, but sterically hindered by a bulky substituent. The experimental results suggest electrostatic effects play the dominant role in the reactions of these substrates.[75]

Treatment of adduct **229c** with DBU causes elimination of the chiral auxiliary and affords 2-carbomethoxynorbornadiene (**231**) in a state of high optical purity.[74] This compound has been employed in racemic form for the preparation of analogues of prostaglandin endoperoxides.[76] This elimination conceivably allows for the nondestructive removal and recovery of the chiral auxiliary. However, no effort was made to determine the overall efficiency of the recovery process.

Utilization of the inverse electron demand strategy has been extended to the reaction of dienyl sulfone **232** and chiral enol ethers **233** derived from optically active secondary alcohols. The reaction produces the bicyclic lactones **234**, without the loss of CO_2, in excellent chemical yield with diastereoselectivities in a range of 0–90%. As seen in Table 13, the best selectivity was obtained with the enol ether prepared from *t*-butyl phenyl carbinol. However, from a practical standpoint, the commercial availability of either (*R*)- or (*S*)-isopropyl phenyl carbinol makes the derived enol ether the reagent of choice.[77]

Table 13. Reactions of Enol Ethers **233** with **232**

R	% Yield	% de	R	% Yield	% de
2-Octyl	>90	0	1-Naph(Me)CH	95	64
endo-2-Bornyl	>90	5	Ph(Me)CH	75	66
Menthyl	89	54	2,4,6-Me₃Ph(Me)CH	>90	80
8-Phenylmenthyl	>90	5	Ph(*i*-Pr)CH	94	84
			Ph(*t*-Bu)CH	90	90

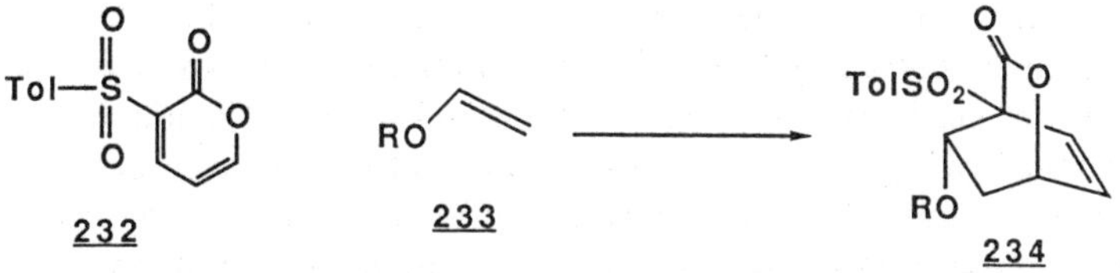

232   **233**   **234**

Asymmetric induction has been reported in the inverse electron demand Diels–Alder reactions of chiral 2-methylene imidizolidines **235** with methyl 2,4-hexadienoates **236**. The adducts have been transformed into optically active cyclohexenones. The reactions suffer, however, from moderate yields and generally low enantiomeric excesses. The best example was for the reaction of methyl-2E,4Z-hexadienoate and imidizolidine **235a** ($R_1 = R_2 = CH_3$; $R_3 = Bz$; $R_4 = H$), which afforded a 53% yield of the cycloadduct **237** (46% ee).[78]

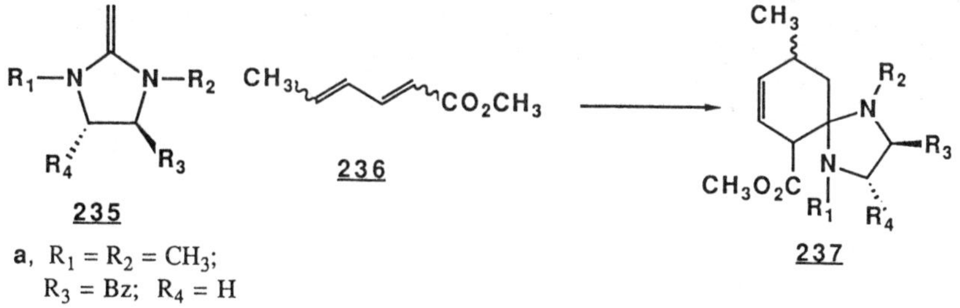

235

a, $R_1 = R_2 = CH_3$;
 $R_3 = Bz$; $R_4 = H$

236

237

The use of a chiral vinyl phosphine oxide, with the chirality at phosphorous, in an enantioselective synthesis of a 17-phosphasteroid system has been communicated by Bodalski *et al.* The uncatalyzed reaction of vinyl naphthalene (**239**) with the vinyl phosphine oxide **238** at 120°C produced a 70% yield of the two diastereomeric adducts **240** and **241** in a ratio of 65:35. The diastereomers could be separated by fractional crystallization and individually processed into the phosphasteroids **242** and **243**.[79]

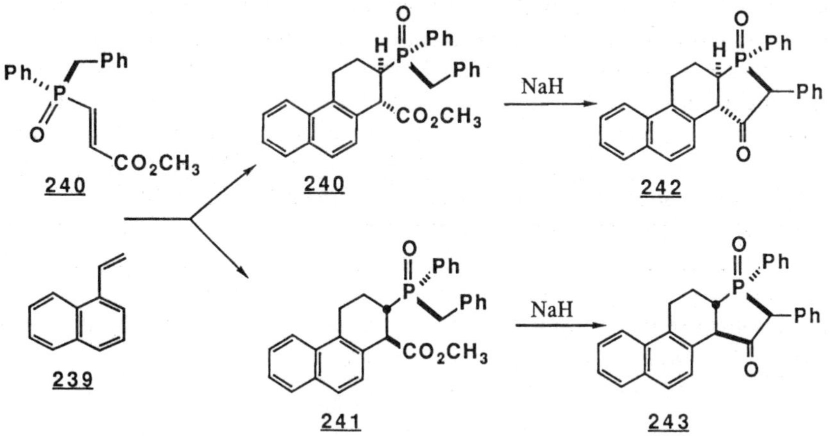

240   **240**   **242**

239

241   **243**

An asymmetric Diels–Alder reaction of an organometallic-based chiral acrylate dienophile equivalent has been reported by Davies and Walker. The reaction of the chiral acyl iron complex **244** with cyclopentadiene in the presence of $ZnCl_2$ produced the cycloadduct **245** in 88% yield as the major diastereomer. The ^{1}H-NMR spectrum also revealed the presence of two other minor diastereomers (21:3:1 ratio). However, the structures of the two minor products were not reported. The absolute configuration and endo selectivity were determined by conversion of adduct **245** into the known iodolactone **246**. A π-stacking interaction between the unsaturated acyl ligand and one of the phenyl rings of the triphenylphosphine ligand has been invoked to explain the shielding of the si-face of the dienophile.[80]

III. CHIRAL HETERODIENOPHILES

Some of the early work with chiral heterodienophiles was delineated by Jurczak in his investigations of the reaction of (R)-(−)-menthyl glyoxylate (**247**) with 1-methoxy-1,3-butadiene (**248**).[81] He found the ratio of the adducts **249** and **250** was dependent on both the solvent and the applied pressure. Table 14 contains a summary of the results obtained from this study. The absolute configuration of the adducts was determined by degradation to dimethyl malate (**251**) and by conversion of the trans isomer **250** into methyl 2,3,6-tri-O-acetyl-4-deoxy-α-D-xylohexopyranoside (**252**). Examination of the results indicates the ratios of **249**:**250** did not vary significantly over the range of conditions examined, and only modest increases in asymmetric induction were observed. In fact, when CH_2Cl_2 was used as solvent, the induction actually decreased with increasing pressure. The most interesting results were obtained with hexane as the solvent, where the absolute configuration of the adduct was reversed as the pressure increased. One final point worth noting was the result of the cycloaddition in $CHCl_3$ as solvent. These conditions produced better inductions at atmospheric pressure than any of the high-pressure conditions.[82]

Table 14. Reaction of **247** with **248** at 20°C

Solvent	Pressure	Configuration	% de	Solvent	Pressure	249 : 250	Configuration	% de
PhCH$_3$	Atmos.	R	3.2	CH$_2$Cl$_2$	2.5 kbar	65 : 35	S	4.4
PhCH$_3$	2.7 kbar	R	4.8	CH$_2$Cl$_2$	7.9 kbar	70 : 30	S	2.4
PhCH$_3$	8.7 kbar	R	8.8	n-Hexane	Atmos.	63 : 37	S	4.0
Ether	Atmos.	R	3.2	n-Hexane	3.1 kbar	66 : 34	S	3.0
Ether	8.5 kbar	R	5.8	n-Hexane	9.3 kbar	70 : 30	R	0.6
Ether	15.5 kbar	R	8.3	n-Hexane	11.1 kbar	71 : 29	R	1.7
CH$_2$Cl$_2$	Atmos.	S	6.2	CHCl$_3$	Atmos.	65 : 35	S	13.9

The 249 : 250 column for the left half:

249 : 250
61 : 39
64 : 36
69 : 31
64 : 36
67 : 33
72 : 28
61 : 39

49

The corresponding reactions of (*R*)-(−)-menthyl glyoxylate (**247**) with 2,3-dimethyl-1,3-butadiene and 1,3-cyclohexadiene produced adducts **253** and **254**. Under optimal conditions, the optical yields were 20.9 and 17.5%, respectively. The chemical yields, however, were low to moderate.[83]

Danishefsky *et al.* have investigated the Lewis acid-mediated "cyclocondensation" reactions of various chiral aldehydes with *trans*-1-methoxy-3-[(trimethylsilyl)oxy]-1,3-butadiene (**255**) (also known as Danishefsky's diene) and its derivatives. The reaction of (*R*)-aldehyde **256** with diene **255** in the presence of $ZnCl_2$ in benzene gave rise to the (5*S*,6*R*)-heptulose **257** in 72% yield. The corresponding reaction with (*S*)-**256** afforded *ent*-**257**, thus paving the way to a variety of difficultly available L-heptoses and L-hexoses. The optical purities of the adducts were ascertained by means of NMR analysis with chiral shift reagents. The (5*S*)-configuration was verified by correlation of **257** with ribonolactone derivative **258**, prepared from 2-deoxyribose. Whether or not chelation is responsible for the excellent Cram selectivities observed is still a matter of conjecture.[84]

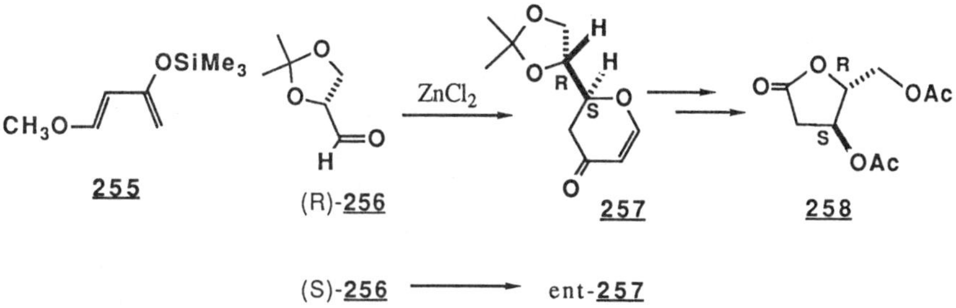

$$(S)\text{-}\underline{256} \longrightarrow ent\text{-}\underline{257}$$

Variations in the solvent system and Lewis acid were found to have a pronounced effect on the stereochemical outcome of these cyclocondensation reactions. The D-galactose-derived aldehyde 259 and diene 260, in the presence of $BF_3 \cdot Et_2O$, produced the mixture of cycloadducts shown below. The stereochemistry of the cis-pyrone 261 was unequivocally established by means of an X-ray crystallographic determination. In the case of aldehyde 259, the cis selectivity was maintained and only the facial selectivity was altered. With other aldehydes studied, the changes were in terms of the cis/trans selectivity. Readers are instructed to peruse the original literature for more dramatic examples of these effects with some achiral and racemic aldehydes.[85]

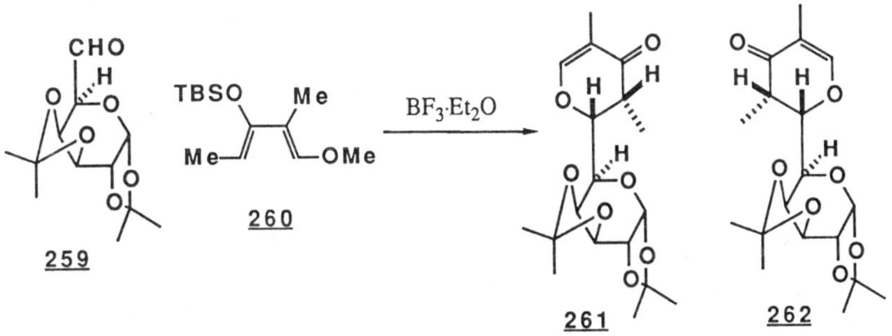

	261	262
CH_2Cl_2	0.9	1.0
$PhCH_3$	4.0	1.0

Two of the cyclocondensation reactions were employed in a synthesis of tunicaminyluracil (268). The route initiated with the reaction of the ribose-derived aldehyde 263 with diene 255. The condensation was catalyzed by the lanthanide shift reagent, $Eu(fod)_3$, and produced the adduct 264 in 85% yield. This adduct was processed into aldehyde 265 in preparation for the second hetero-Diels–Alder reaction. Aldehyde 265 was reacted with diene 266 in the presence of the novel catalyst system, $Ce(OAc)_3$–$BF_3 \cdot Et_2O$, to afford the cycloadduct 267 as a single isomer. A series of transformations converted **267** into **268**, which was identical in all respects to a compound prepared from tunicamycin (**269**).[86]

R= -CH=CH-(CH$_2$)$_n$-CHMe$_2$
n= 7, 8, 9, 10, 11
R= -CH=CH-(CH$_2$)$_n$-CH$_3$
n= 10, 11, 12, 13

Recently, the ionophore zincophorin (**277**) has succumbed to total synthesis by the Danishefsky group. Again, two hetero-Diels–Alder reactions were used to assemble key portions of the natural product. The magnesium bromide-mediated reaction of diene **270** and aldehyde **271** furnished compound **272** as a 7:1 trans/cis mixture presumably via chelation control. Conversion of **272** to aldehyde **273** set the stage for the second cyclocondensation reaction with the 4*E* diene **274**. Under the influence of $BF_3 \cdot Et_2O$, the reaction gave rise to the *trans*-pyrone **275**, which was subsequently manipulated into (+)-zincophorin methyl ester (**276**).[87]

The hetero-Diels–Alder protocol was also employed in a total synthesis of the aglycone of avermectin A_{1a}. The D-glucal-derived aldehyde **278** was reacted with diene **279** in the presence of $MgBr_2$ in CH_2Cl_2 to provide a 75% yield of a 3.5–5.0:1 ratio of adducts with the major product being the pyrone **280**. This was ultimately processed into the desired aglycone **281**.[88]

An asymmetric synthesis of *threo*-β-hydroxy-L-glutamic acid (**284**) has been achieved by Garner via the Lewis acid-catalyzed hetero-Diels–Alder reaction of the penaldic acid equivalent **282** with Danishefsky's diene (**255**). The D-serine-derived aldehyde **282** and diene **255** in the presence of $ZnCl_2$ provided the *threo*-pyrone **283** in 70% yield as a > 9:1 mixture of isomers. The major isomer was converted in a short sequence of reactions to the glutamic acid **284**.[89a] Recent investigations have resulted in an increase in the selectivity (60:1) and shown the threo selectivity is directly proportional to the amount of $ZnCl_2$ and inversely proportional to the solvent polarity.[89b]

Jurczak has utilized the high pressure technique in a total synthesis of 6-epi-D-purpurosamine B (**288**). The L-alanine-derived aldehyde **285** was condensed with methoxy butadiene in the presence of $Eu(fod)_3$ at 50°C under 20 kbar of pressure to yield a mixutre of four possible diastereomers in 70%. The mixture was isomerized under acidic conditions to provide a 2:1 mixture of adducts **286** and **287**, respectively. The two adducts were separated and cycloadduct **286** was transformed in a series of steps to the final product **288**.[90]

Nitsch and Kresze have reported the asymmetric Diels–Alder cycloaddition of the α-chloronitroso compound **289** with 2,4-hexadiene in CH$_3$OH. This gave product **291** in 83% yield. The chiral auxiliary was lost during the cycloaddition reaction via the hydrolysis of the isomeric iminium salt of the initially formed adduct **290**. This allows for the convenient recycling of the auxiliary. Degradation of the adduct to O-acetyllactic acid (**292**) established an optical purity of 39% and the (S)-configuration at C-6 in the adduct.[91]

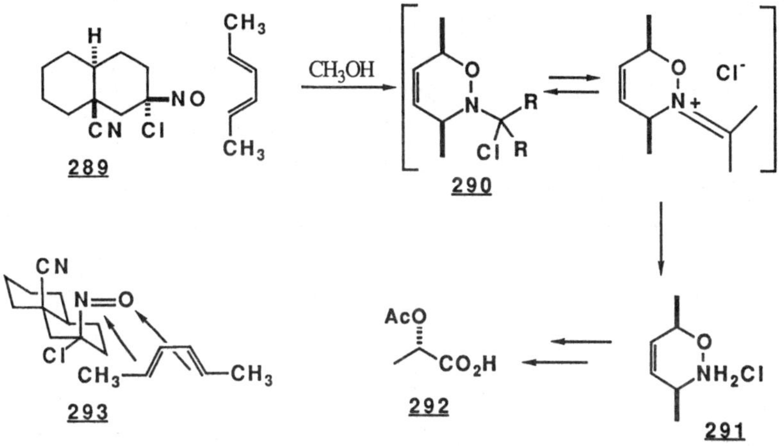

The results have been explained in terms of the transition state model **293**, shown above, with the diene approaching the NO group as indicated. The position of the NO group relative to the decalin system is based on the analysis of the X-ray crystal structure of **289**.

Subsequent to these studies, Kresze and co-workers prepared the α-chloronitroso compound **295**, derived from epiandrosterone (**294**). Reaction of **295** with 1,3-cyclohexadiene produces adduct **296** with the (1R,4S)-configuration in 69% yield with an enantiomeric excess of > 95%. Again, the chiral auxiliary is lost during the cycloaddition by cleavage of the intermediate iminium salt. The configurations were established by degradation and correlation with a known glutaric acid derivative.[92]

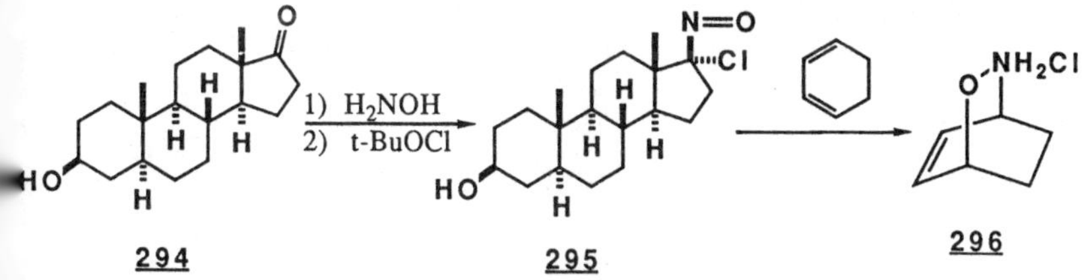

approach of the diene to the nitroso compound in transition state **297**, as depicted below, has been proposed to explain the observed configurations in the final product.

297

A number of communications describing the cycloadditions of the chiral triazolinones **298–300** have been reported by Paquette and co-workers.[93] These highly reactive heterodienophiles participate well in the Diels–Alder reactions, but do not include useful levels of enantioselection.[94] They have, however, proved useful for the nondestructive resolution of nonobviously resolvable compounds. Triazolinone **300**, derived from (+)-camphor, has achieved the greatest success in this regard and has consequently received the

298   299   300

most attention. The overall resolution process is illustrated for 1,2,3-trimethylcyclooctatetraene (**301**). The adducts **302** and **303** could be separated by crystallization and individually processed into the highly enantiomerically enriched cyclooctatetraenes (+)- and (−)-**301**.[95] This strategy has been successfully implemented in the syntheses of a number of substituted cyclooctatetraenes,[96] propellanes **304** and **305**,[97] (+)-methylsemibullvalene (**306**),[98] and (+)-9-nortwistbrendane (**307**).[99]

Complete asymmetric inductions have been achieved in the cycloaddition reactions of the chiral sulfines **308**. Reaction of these sulfines with 2,3-dimethyl-1,3-butadiene afforded adducts **309**, which within the limits of detection by ^{1}H-NMR and HPLC analysis were formed with complete asymmetric induction. An X-ray analysis of adduct **309a** revealed the absolute configurations at the chlorine-bearing carbon and the sulfur to be R. It also demonstrated that the stereochemistry of the starting sulfine, shown by X-ray analysis to be Z, had been retained in the product. The results seem to indicate the diene has added to the $C_{si}S_{re}$-face of the rotomer with the chlorine syn to C-7, thus avoiding the C-2 carbonyl.[100]

Zwanenburg and co-workers have also disclosed the results of the asymmetric Diels–Alder reactions of the sulfines **310**, derived from S-proline, with 2,3-dimethyl-1,3-butadiene. The results obtained show only moderate asymmetric induction, and in some cases no induction, with these hetero-dienophiles.[101]

a, R_1= CH_3; R_2= Cl	21% de
b, R_1= CH_3Mes; R_2= Cl	38% de
c, R_1= CPh_3; R_2= Ph	0% de
d, R_1= R_2= CH_2Ph	36% de

Two independent reports by Whitesell and co-workers and Weinreb and co-workers have described enantioselective Diels–Alder reactions with chiral *N*-sulfinylcarbamates. The Texas group reacted sulfinylcarbamate **313** with (*E,E*)-hexa-2,4-diene under thermal and Lewis acid-catalyzed conditions. The uncatalyzed reaction produced a mixture of the four diastereomers **314–317** in a ratio of 2:1:2.5:9. In the presence of SnCl$_4$, only diastereomer **315**, the minor product of the thermal reaction, was obtained in 42% isolated yield. The structure of **315** was confirmed by a single crystal X-ray structure determination. When **313** was reacted with the unsymmetrical diene *E*-piperylene, under the influence of SnCl$_4$, only one regio- and stereoisomeric adduct **318** was formed to the extent of at least 98:2.[102]

In the work conducted by the Penn State group, *N*-sulfinylcarbamate **313** was reacted with 1,3-cyclohexadiene under both thermal and Lewis acid-mediated conditions. The thermal process produced an 80% yield of a mixture of four diastereomers in a ratio of 2:4:1:3. The structural identity of the diastereomers was not reported. When the reaction was conducted at − 50°C in the presence of TiCl$_4$, a 77% yield of adduct **319** was obtained as a 9:1 mixture of epimers at sulfur. The absolute configuration of **319** was determined to be 3*S*,6*R* by conversion of the cycloadduct to the known cyclic carbamate **320**.[103]

Also investigated was the chiral sulfinylcarbamate **321**, with the auxiliary derived from (+)-camphor. The uncatalyzed reaction with cyclohexadiene again produced an unidentified mixture of four diastereomers. The TiCl$_4$-catalyzed reaction gave exclusively one adduct **322** whose configuration at sulfur could not be determined. The absolute configuration at carbon was established as 3*S*,-6*R* by conversion to carbamate **320**.[103]

Larsen and Grieco have performed an asymmetric aza Diels–Alder reaction in aqueous solution. The reaction of (−)-α-methylbenzylamine

hydrochloride, aqueous formaldehyde, and cyclopentadiene at 0°C produced a 4:1 mixture of separable diastereomers **323** and **324** in 86% yield.[104]

IV. CHIRAL DIENES

In comparison to the amount of work done with chiral dienophiles
relatively little has been done with chiral dienes. The first report dealing with
the Diels–Alder reaction of a chiral diene was disclosed in 1974 by David *et
al.* The cycloaddition of the carbohydrate-based diene **325** with *n*-butyl
glyoxylate produced a mixture of all four possible diastereomers **326–329**.[105]
The dienyl sugars **330–333** have also been investigated in the cycloaddition
reaction with the glyoxylate ester. The results of these studies are shown in
Table 15.[106] This methodology has been applied in synthetic approaches to
the blood group A antigenic determinant[107] and kasugamycin.[108]

The glyoxylates do not react with cis-dienes but this problem can be overcome by using diethyl mesoxalate (**334**) as the heterodienophile. The resultant malonic ester cycloadducts can then be selectively decarboxylated using NaCl in wet HMPA.[109,110]

Table 15. Cycloadditions of **325**, **330–333** with Butyl Glyoxylate

Diene	α-D	β-D	α-L	β-L	% endo	% topface	% bottomface
325	13	18	9	60	78	27	73
330	44	4	0	52	56	4	96
331	48	18	0	34	52	18	82
332	1.1	50.3	46.5	2.1	52.4	96.8	3.2
333	20	25	47	8	33	72	28

The sugar-based diene **335** was utilized in a synthesis of (+)-4-demethoxy-daunomycinone (**338**). Reaction of **335** with oxirane **336** produced a 2.5:1 mixture of cycloadducts with the major diastereomer being **337**, which was transformed into **338**. This diene has also been shown to react with other dienophiles to afford reasonable yields of cycloadducts with diastereomeric ratios ranging from 3 to 8:1.[111]

Trost *et al.* employed the (*S*)-*O*-methylmandeloxydiene **339** in an enantio-selective synthesis of ibogamine (**341**). The cycloaddition of **339** with acrolein in the presence of $BF_3 \cdot Et_2O$ derived in 92% yield adduct **340**, with 80% of the adduct having the (3*R*,4*S*,6*R*)-configuration and 20% possessing the (3*S*,4*R*,6*S*)-configuration. The adduct was ultimately transformed into (+)-ibogamine (60% ee).[112]

Reaction of (*S*)-mandeloxybutadiene **70** with acrolein under similar con ditions ($BF_3 \cdot Et_2O$ catalysis) provided the adduct **342** in 90% yield as an 82 18 mixture of the (3*R*,4*S*)- and (3*S*,4*R*)-isomers, respectively. Interestingly the reaction of (*S*)-**70** with juglone (**343**) in the presence of 1.6 equivalents o boron triacetate furnished cycloadduct **344** as a single diastereomer.[113]

In all of these reactions, the absolute configuration of the final product has been explained by proposing the reactive conformer for the diene to b best represented by **345**. The alternative conformation **346** was judged les likely on the basis of steric interactions. In conformer **346**, the projection c the large group toward the diene results in a severe nonbonded interaction. The corresponding nonbonded interaction between the small group and the diene in the reactive conformer is not as severe. Thus, the aromatic ring serves to guide the dienophile to one of the diene faces. It has also been suggested that the π-stacking interaction between the diene and the aromatic ring becomes increasingly more important as the charge transfer between the diene and dienophile increases. This is evidenced by the increase in asymmetric induction in going from acrolein to juglone as the dienophile, where juglone is expected to form stronger charge-transfer complexes. Further indication of the importance of the π-stacking interaction can be found in the passive behavior of diene 347 toward cycloaddition, which has been ascribed to the sandwiching of the diene between the aromatic rings.[113]

The diene 349, available from the chiral enedione 348, has been employed by Cohen in an asymmetric synthesis of D-homo steroids. The AlCl$_3$-mediated cycloaddition of 349 with 2,6-dimethyl-*p*-benzoquinone (350) produced a 2:1 mixture of adducts 351 and 352. The major isomer could be separated from the minor one by fractional crystallization. The overall process afforded 351 in 25–30% yield from the enedione 348.[114]

The effect of high pressure on the asymmetric induction in the Diels–Alder reaction of chiral 2,4-pentadienoic acid derivatives 353 with *p*-benzoquinone

354 has been investigated by Dauben and Bunce. Some of the results of his study are shown in Table 16. The yields for the process were good, but the enantiomeric excesses were moderate at best.[115]

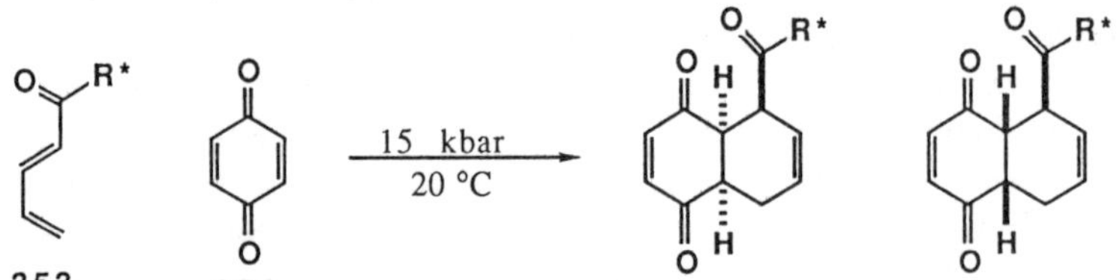

353   **354**

Table 16. Asymmetric Induction in High-Pressure Reactions of **353** and **354**

R*	% Yield	% ee	R*	% Yield	% ee
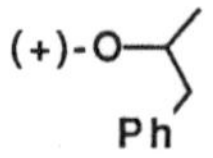 (+)-O— Ph	98	2	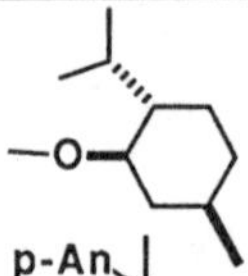 —O— p-An	58	36
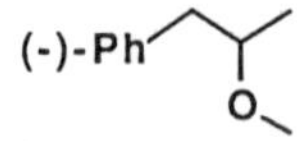 (-)-Ph— O	98	6	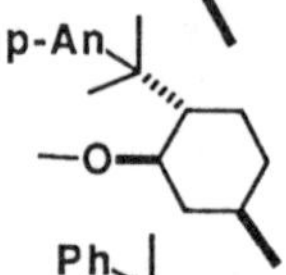 —O— Ph	60	46
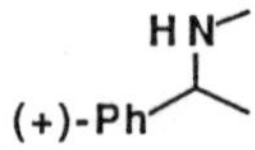 HN— (+)-Ph	94	14	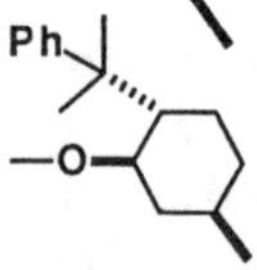 —O—	62	50

Methyl diformylcamphanate **355** undergoes an inverse electron-demand Diels–Alder reaction with dihydrofuran at 20°C to yield a mixture of diastereomeric adducts **356** and **357**. The pairs **356a/357a** and **356b/357b** were obtained in a 1:2 ratio. A diastereomeric excess of 52% was achieved on the basis of the 1:3.2 ratio of **356b** to **357b**, the latter of which can be isolated in 41% yield by recrystallization.[116]

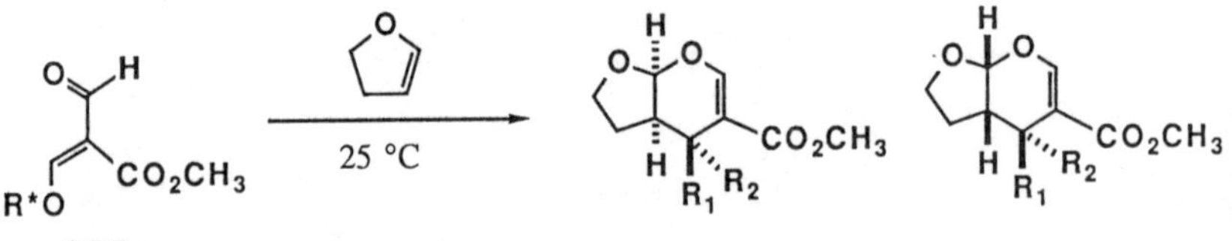

355

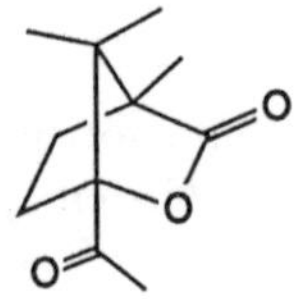

356a, R_1= OR*; R_2=H   **357a**, R_1= OR*; R_2=H

356b, R_1= H; R_2=OR*   **357b**, R_1= H; R_2=OR*

R*=

A similar inverse type hetero-Diels–Alder reaction has been applied to the synthesis of C-arylglycosides. The reaction of the O-methylmandeloyl 1-oxa diene **358** was reacted at 5 kbar with styrene **359** to produce exclusively the endo adduct **360** as an unspecified 3:1 mixture of diastereomers. The adduct was converted in a short sequence into the C-aryl-2-deoxy-β-glucopyrano-side **361**.[117]

Ito *et al.* have extended their fluoride anion-induced 1,4-elimination of o-[1-(trimethylsilyl)alkyl]benzyltrimethylammonium halides for the mild and efficient generation of o-quinodimethanes to the chiral 2-[o-[1-(trimethylsilyl)-alkylphenyl]-3,3-dimethyloxazolidinium salt **362**, which is derived from (−)-ephedrine. Treatment of **362** with CsF in the presence of methyl acrylate at 0°C resulted in a 92% yield of adduct **363** as a 2 : 1 mixture of diastereomers. The major diastereomer was separated and converted to the known carb-oxylic acid **364**. Comparison of its rotation with the known rotation of **364** established the absolute configuration as R and an enantiomeric excess of 67%. This translates to a (1R, 2R)-configuration for the major diastereomer obtained in the cycloaddition to the o-quinodimethane.[118]

Again π-stacking interactions have been invoked to explain the shielding of one of the faces of diene. It has been postulated that of the two possible conformations, **365** and **366**, the nonbonded interactions in **365** between the benzylic and aromatic hydrogens render this conformation less favorable. However, no mention was made concerning the interactions of the other obviously much larger benzylic substituent.

365   **366**

This conformational analysis and the importance of the π-stacking interactions to rationalize the observed induction in the cycloaddition of the
intermediate o-quinodimethane have been challenged by Charlton.[119] In a
systematic study, a number of chiral alkoxy groups were examined to assess
the effect of the position of the phenyl group and the position of the chiral
center on the degree of asymmetric induction. The alkoxy benzosulfones **367**
were employed as the progenitors of the o-quinodimethanes. The results of
the reactions of the benzosulfanes **367** with dimethyl fumarate presented in
Table 17 reveal the (R)-sec-phenylethoxy auxiliary to be the most effective in
differentiating the diastereotopic faces of the diene.

367   **368**   **369**

Table 17. Asymmetric Diels–Alder Reactions of **367**

R*		% Yield	368:369	R*		% Yield	368:369
a	CH₃ —CH—Ph H	85	2.8 : 1	d	CH₃ -CH₂CH—Ph H	63	1 : 1
b	CH₃ —CH—CH₂Ph H	77	1.9 : 1	e	CH₃ —CH—Cyclohexyl H	56	1 : 1
c	CH₃ —CH—CH₂CH₂Ph H	79	1 : 1				

The absolute configuration of the adducts was determined by trapping the *o*-quinodimethane with methyl acrylate and transformation of the adduct **370** to the known acid **364**. The acid was assigned the (*S*)-configuration on the basis of its negative rotation. The adduct **370** was therefore assigned the (1'*R*, 1*S*,2*S*)-configuration.

In Charlton's analysis of the problem, an attempt to apply a π-stacking arrangement to justify the observed induction resulted in some inconsistencies. The possible conformations for the π-stacking interactions are depicted by **371** and **372**. The apparently more favorable conformer **372** would have resulted in the (*R*)-configuration for the cycloadducts. Also, in those cases where π-stacking would be more sterically possible, no inductions were observed. For this reason, conformation **373** was deemed the more likely candidate to explain the induction on steric grounds. However, the strength of this argument was weakened by the case in which no induction was observed when the phenyl **367a** was replaced by a cyclohexyl group.

Winterfeldt and co-workers have examined the use of cycloadducts prepared from steroidal derived dienes as templates for diastereoselective transformations. After serving its purpose, the diene could be regenerated and pure enantiomers expelled in a retro-Diels–Alder reaction. The first attempt at deploying this strategy involved the cycloaddition of ergosterol acetate (**374**) with propargyl aldehyde (**375a**). The expected adduct **376** was obtained only when the reaction was conducted in refluxing CH_2Cl_2 with WCl_6 as the Lewis acid (no yield given). If the reaction were run in refluxing toluene or adduct **376** was heated in toluene, the aromatic aldehyde **377** was formed in 82% yield. The retro-Diels–Alder reaction had not ejected the acetylenic aldehyde, but had fragmented the hydrophenanthrene system to produce the ansa-steroid **377**. The structure of the adduct was unequivocally established by means of an X-ray structure determination.[120]

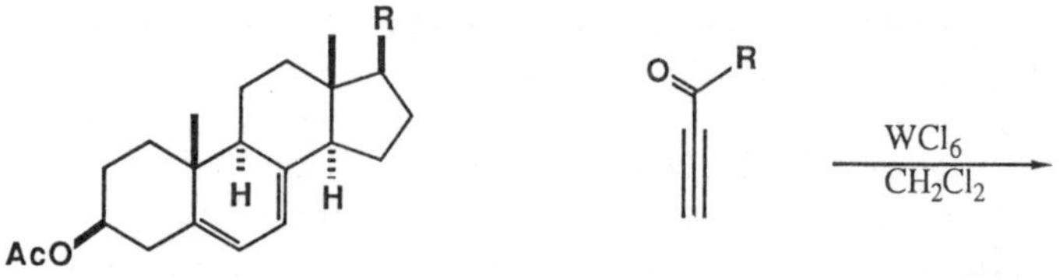

374   375a, R= H

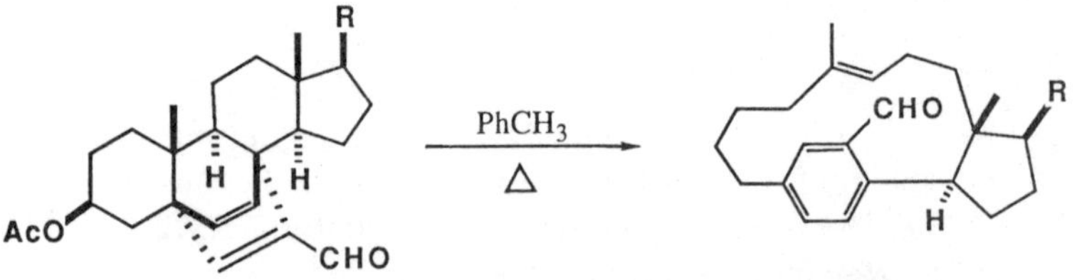

376   377

In a second generation approach designed to avoid the ansa-steroid formation, the steroid-derived cyclopentadiene **378** was examined. Cyclo-addition of **378** with propargyl aldehyde (**375a**) or butynone (**375b**) afforded regio- and stereoselectively adducts **379a** and **379b** in > 90% yield. Adduct **379b** could be manipulated by cuprate addition, ketone reproduction, acetylation, and retro-Diels–Alder reaction (200–220°C) into enantiomeri-cally pure (> 90%) acetate **380**.[121]

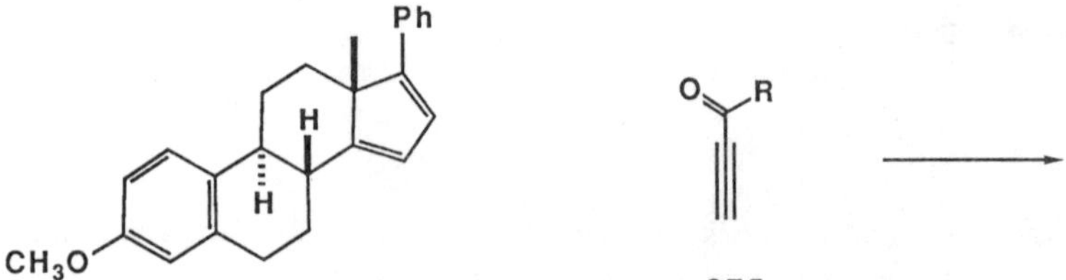

378   375

                                       a, R=H
                                       b, R= CH$_3$

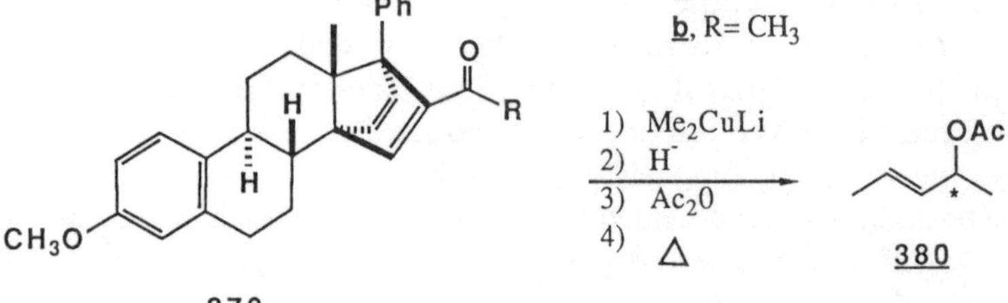

379

1) Me$_2$CuLi
2) H$^-$
3) Ac$_2$O
4) Δ

380

Kozikowski *et al.* have examined the π-facial selectivity of the L-threonine derived diene **381** in a synthetic program that has culminated in the total synthesis of actinobolin (**107**). Reaction of **381** with methyl propiolate produced a mixture of diastereomers **382** and **383**, with the ratio being dependent on reaction conditions. The ratio was found to vary from 3:1 (110°C, 24 h, PhH, 85%) to 1.7:1 (220°C, 3 h, neat, 87%). The assignment of the structures to the individual diastereomers was made on the basis of an X-ray structure of lactone **384**, derived from the major diastereomer **382**. Unfortunately, the stereochemistry at C-10 was opposite of what was required for the synthesis of actinobolin. Lactone **384** was converted in a sequence of transformations to (+)-5,6,10-tri-(epi)-actinobolin (tri-epi-**107**).[122a] The minor diastereomer **383** has been transformed along similar lines into **107**.[122b,c]

The following transition state picture **385** has been proposed to account for the formation of **382** as the major diastereomer. The carbomethoxy group in the linear acetylene encounters less steric interference with the amido group in the outside position as opposed to having the amido group in the inside position. Electron withdrawal from the diene by the nitrogen substituent is also minimized by having the heteroatom positioned anti to the newly forming bond in this transition state and therefore approach of the dienophile to the diene face opposite the silyloxy ethyl moiety is favored.[122c]

A formal total synthesis of the Inhoffen–Lythgoe diol (391), an intermediate in the syntheses of vitamin D_3 and related metabolites, has been communicated by Grieco et al.[123] The aqueous Diels–Alder reaction of methacrolein (386) with the diene 387a gave rise to a 70% yield of diastereomer 388a along with 15% of 389a. When the methyl ester 387b was reacted with neat methacrolein at 55°C, only a 10% yield of a 1:1 mixture of 388b:389b was realized. The structures were deduced on the basis of an X-ray analysis of the minor diastereomer. Adduct 388a was transformed into 390 which has previously been converted into 391.[123]

386   387
           a, R = Na
           b, R = CH₃

388   389

390   391

The cyclocondensation reaction of aldehydes with activated dienes has been investigated with the l- and d-menthyloxy auxiliaries attached to the diene. When the l-menthyloxy dienes 392 were reacted with benzaldehyde in the presence of Eu(fod)₃, only moderate selectivities for the "D-pyranose" products 394 over the L-isomers 395 were observed. The use of the mild lanthanide catalyst allowed the isolation of the otherwise labile primary adducts. The extent of the inductions was determined by conversion of the cycloadducts to the dihydropyranones 396 with trifluoroacetic acid (TFA). Under identical conditions the reactions of the d-menthyloxy dienes 393 expressed an equal and opposite propensity for the formation of the "L-pyranose" products 395.[124b,c]

392, R*= l-Menthyl

393, R*= d-Menthyl

394

395

D-**396**

a, X= Y= H   27% ee

b, X= CH$_3$; Y= H   10% ee

c, X= OAc; Y= H   10% ee

d, X= Y= CH$_3$ 2% ee

V. CHIRAL CATALYSTS

Discovery of the effectiveness of the lanthanide metals to catalyze the cyclocondensation reaction[124a] made possible the examination of the chiral shift reagent (+)-Eu(hfc)$_3$ as a catalyst that might bestow some enantio-selectivity on the overall process. The reactions of dienes **397** with benz-aldehyde in conjunction with (+)-Eu(hfc)$_3$ produced, after trifluoroacetic acid treatment, adducts D-**396** and L-**396** with only modest enantiotopic face differentiation. The data reveal this chiral catalyst consistently affords an excess of the L-dihydropyrones.[124,125]

397

398

L-**396**

a, X= Y= H   38% ee   **b**, X= CH$_3$; Y= H   40% ee

c, X= OAc; Y= H   34% ee   **d**, X= Y= CH$_3$ 42% ee

A remarkable interactivity between the chiral catalysts and the chiral auxiliaries was discovered in the hetero-Diels–Alder reaction of the activated dienes with the aldehyde dienophiles.[124b,c] Cycloaddition of the *d*-menthyl-oxy dienes **393** with benzaldehyde in the presence of (+)-Eu(hfc)$_3$ followed by quenching with Et$_3$N/CH$_3$OH produced the isomers **399** and **400** in the ratios indicated. These results differ very little from those obtained with the achiral Eu(fod)$_3$ catalyst. In this case, there appears to be little interactivity between the chiral elements. This is contrasted with the results obtained in the (+)-Eu(hfc)$_3$-catalyzed reaction of *l*-menthyloxy diene **392** and benz-aldehyde. The combination of the modestly L-selective catalyst with the normally D-selective diene **392** produced the highest ratios of L/D-pyranoses. It was deemed highly unlikely that the increase in facial selectivity was the result of double diastereoselection. In this instance, the facial bias of the auxiliary was inverted on interaction with the chiral catalyst.[124c]

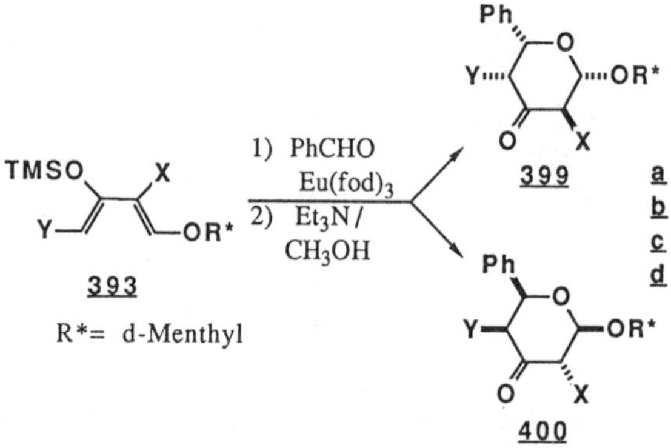

X	Y	399:400
a H	H	63:37
b CH$_3$	H	59:41
c OAc	H	59:41
d CH$_3$	CH$_3$	51:49

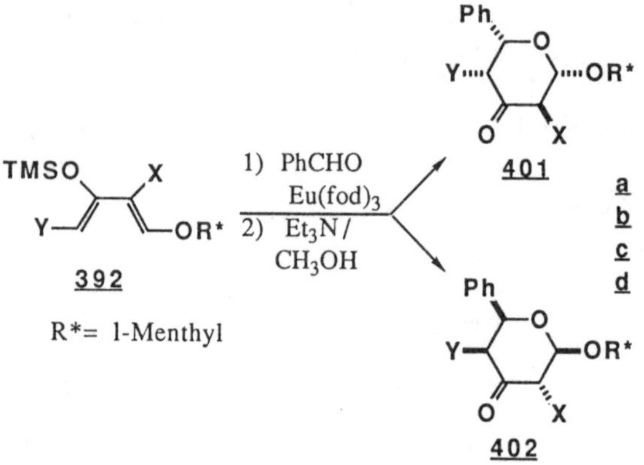

R*= d-Menthyl

X	Y	401:402
a H	H	75:25
b CH$_3$	H	98:2
c OAc	H	93:7
d CH$_3$	CH$_3$	87:13

R*= l-Menthyl

In the hope of obtaining even higher diastereofacial selectivities in the cyclocondensation process, an examination of diene **403** containing the chiral auxiliary *l*-8-phenylmenthol was undertaken. Reaction of **403** with benzaldehyde in the presence of (+)-Eu(hfc)$_3$ provided L-pyranose **404** as a 25:1 mixture with its D-isomer. This adduct was transformed in a series of steps to the unnatural L-glucose (**405**).[124c]

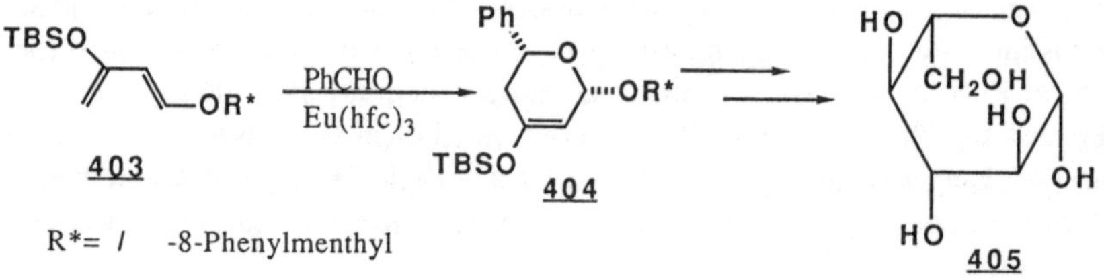

R*= *l* -8-Phenylmenthyl

This technology was also applied to the preparation of the L-oleandrose residues of avermectin A_{1a}. Diene **403** was reacted in a lanthanide-catalyzed condensation with acetaldehyde and the adduct processed into L-dihydropyrone **406**. This was subsequently converted into the unsaturated disaccharide derivative **407**, which was then coupled with the aglycone **281** (*vide supra*). A simple three step sequence resulted in the synthesis of avermectin A_{1a} (**408**).[88b]

This was not the first application of chiral catalysis in the asymmetric Diels–Alder reaction. The first reported use of chiral catalyst was in 1976 when the reaction of methyl acylate and cyclopentadiene catalyzed by the menthyl ethyl ether–BF_3 complex produced the desired adduct in 3.3% enantiomeric excess.[126]

Chiral alkoxyaluminum dichlorides **409–411** have been tested by Koga and co-workers in the Diels–Alder reaction of various acrylates with cyclopentadiene. The most effective combination was with catalyst **409** and methacrolein. These reportedly produced adduct **412** with an enantiomeric excess of 72%.[127] Some skepticism of these results has recently surfaced because Oppolzer has only been able to attain an enantiomeric excess of 55% under the same conditions with catalyst **409**.[8a] These data have since been revised in a report from Koga's group. The enantiomeric excess of 72% has been readjusted to 57%.[127b] The stereochemistry has been rationalized via the transition state **413**. This is based on the most stable conformation of menthyl acrylate that has been calculated by Houk and co-workers (*vide supra*).[24]

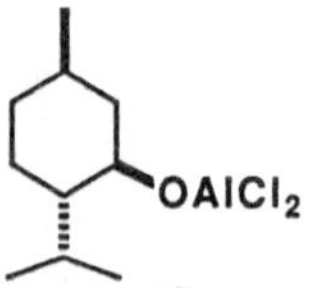

409 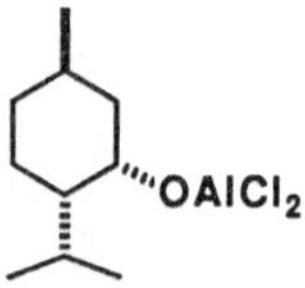   410    411 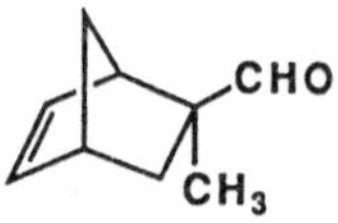   412

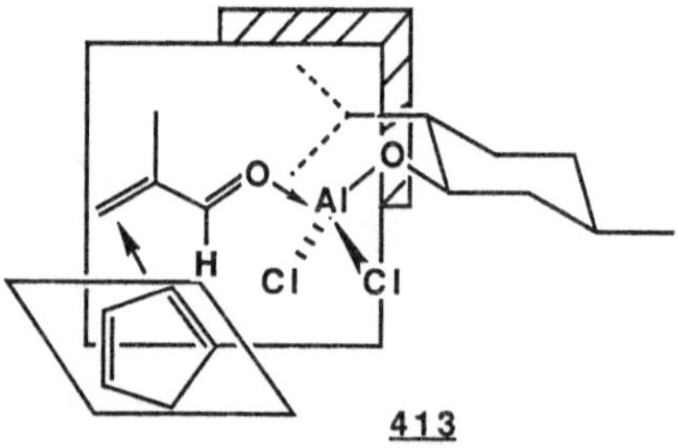

413

Catalyst **409** has been shown to catalyze the reaction between acrylonitrile and 1,3-butadiene to produce adduct **414** with a 2.9% optical purity. In addition, the reaction of methyl acrylate and cyclopentadiene yields **415** in 2.9% optical purity under the influence of this catalyst. It has also been shown to be the most effective catalyst of those surveyed for inducing asymmetry into the product of the copolymerization of acrylonitrile and benzofuran.[128]

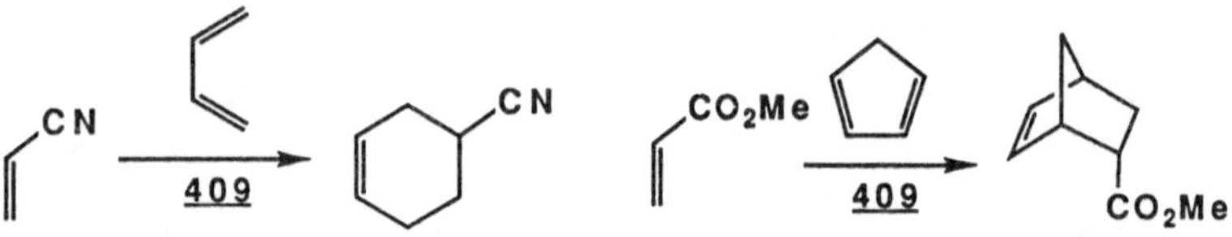

414   415

A comparison of **409** and Eu(hfc)$_3$ for their ability to catalyze the hetero-Diels–Alder reaction between butyl glyoxylate and methoxy butadiene has shown the Europium catalyst to be superior.[129]

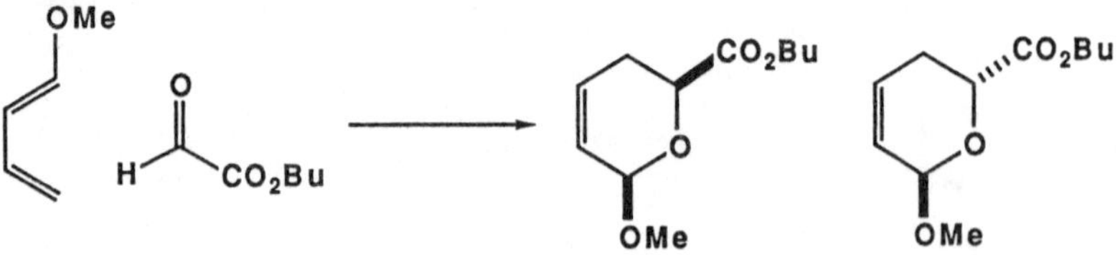

Cat. = **409**	17% (16% ee)	9% (3% ee)
Cat. = Eu(hfc)$_3$	19% (64% ee)	79% (39% ee)

The mono- and diisopinocamphenylhalogen boranes **416–419** have been employed in the reaction of methacrolein and cyclopentadiene. The use of borane **416** results in the formation of adduct **412** (exo:endo; 90:10) with an enantiomeric excess of 28.5%.[130]

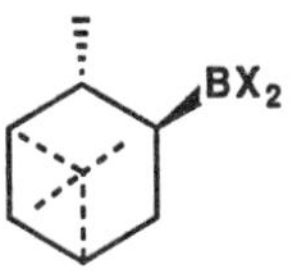

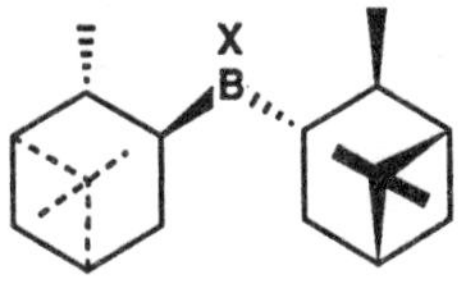

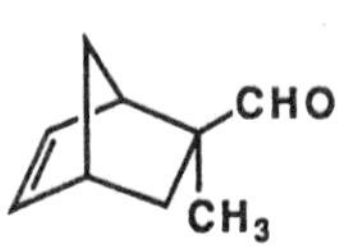

416, X = Cl

417, X = F

418, X = Cl

419, X = Br

412

A few chiral titanium reagents have been prepared from diols **420–422**. The catalyst **423**, prepared from diol **422**, has been shown to be the most effective in the reactions between the oxazolidinones **424a–e** and cyclopentadiene. The results are tabulated in Table 18.

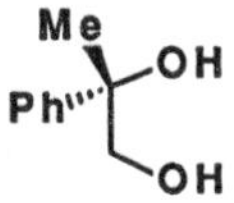

420

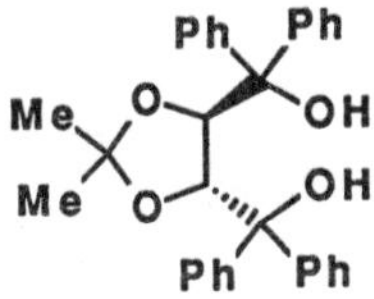

421

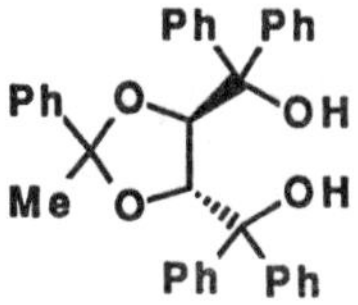

422

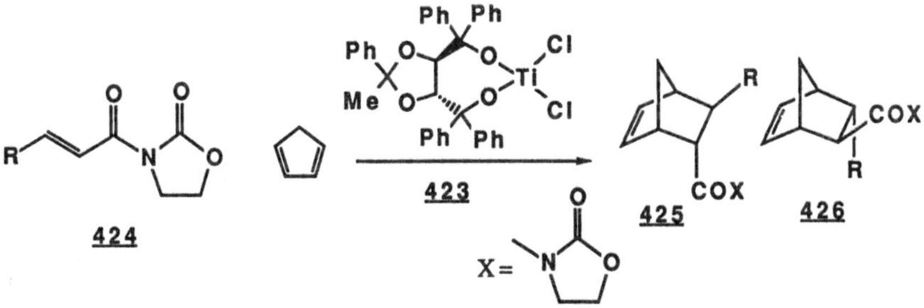

424

423

425

426

$$X = \text{(oxazolidinone)}$$

Table 18. Asymmetric Diels–Alder Reactions of **424** Catalyzed by **423**

424	% Yield	endo : exo	% ee of 425
a R = Me	93	90 : 10	92
b R = H	69	86 : 14	38
c R = Ph	97	92 : 8	81
d R = *n*-Pr	82	90 : 10	90
e R  = (Me-alkenyl)	77	92 : 8	82

Catalyst **423** also catalyzed the reaction of **424a** and isoprene to produce cycloadduct **427** of undetermined absolute configuration in 92% ee.[131a] A substituent in the β-position of the acrylate appears to be essential for reasonable induction. The enantiomeric excess of the adduct obtained from **422b** and cyclopentadiene can be raised to 64% by using 4 Å molecular sieves. The use of sieves has no beneficial effect on the reactions with the other dienophiles.[131b]

In a similar study catalysts were prepared from diols **428**–**433** and examined for their ability to control π-face stereodifferentiation in catalyzed Diels–Alder reactions of the oxazolidinones **424a** and **424b**. These gave excellent results with **424a** and cyclopentadiene as can be seen in Table 19. The results with dienophile **424b** were less encouraging.[132]

Table 19. Asymmetric Induction in the Diels–Alder Reactions of **424** Catalyzed by Chiral Lewis Acids

Lewis acid	Ligand	**424** : Lewis acid : ligand	% Yield	endo : exo	% de
EtAlCl$_2$	**428a**	1 : 2 : 1	89	75 : 25	94
AlCl$_3$	**428b**	1 : 1 : 1	55	88 : 12	92
TiCl$_4$	**428b**	1 : 1 : 1	86	93 : 7	96
EtAlCl$_2$	**429**	1 : 2 : 1	73	73 : 27	92
EtAlCl$_2$	**430a**	1 : 2 : 1	92	76 : 24	95
TiCl$_4$	**430b**	1 : 1 : 1	99	94 : 6	> 98
EtAlCl$_2$	**431**	1 : 2 : 1	90	70 : 30	93
EtAlCl$_2$	**432**	1 : 2 : 1	71	73 : 27	91
EtAlCl$_2$	**433**	1 : 1 : 0.5	71	84 : 16	94

A highly effective chiral Lewis acid that produces high levels of asymmetric induction has been cleverly and rationally designed by Kelly et al.[133] The catalyst system is boron based and utilizes the bidentate ligand **430a**, which possesses C_2 symmetry, in combination with peri-hydroxyquinones. The hydroxy group of the quinone functions as a second ligand for the Lewis acid and helps restrict the conformational mobility in the resulting complex. Thus, reaction of BH_3, AcOH, and **430a**, followed by addition of juglone (**343**), led to the formation of the intermediate complex **435**. Addition of dienes **436** and **437** afforded 70–90% yields of the cycloadducts **438** and **439** in >98% ee.

This catalyst system has been applied with comparable success in the cycloaddition of quinone **440** with diene **441** to provide the adduct **442**. Compound **442** has been applied to a total synthesis of (−)-bostrycin (**443**).

Subsequent to Kelly's report, Yamamoto and co-workers described the use of a similar catalyst system with derivatives of (R,R)-(+)-tartaric acid as the bidentate ligands. The best results were obtained with the diaryl tartaramide **444** derived from m-toluidine. The reaction of **343** with this catalyst system and diene **445** afforded a 73% yield of adduct **446** in 92% ee.[134]

The same Japanese group has communicated the use of the binaphthol-aluminum catalysts **447** and **448** in the asymmetric hetero-Diels–Alder reaction. The reaction of benzaldehyde and diene **270** under the influence of 10 mol% of **447a** afforded, after TFA treatment, L-dihydropyrone L-**396d** (77% yield, 95% ee). A small amount (7%) of the trans isomer was also isolated. The use of **447b** resulted in an increased yield (93%), cis:trans ratio (30:1), and enantiomeric excess (97%).[135]

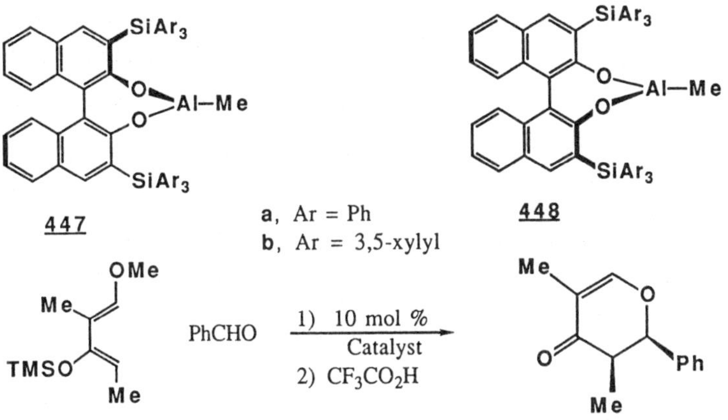

447

a, Ar = Ph
b, Ar = 3,5-xylyl

448

270

PhCHO

1) 10 mol %
   Catalyst
2) CF₃CO₂H

L-396d

VI. ASYMMETRIC INTRAMOLECULAR DIELS–ALDER REACTIONS

An examination of the effect of the optically active Lewis acids (*l*-menthyl-oxy)aluminum dichloride **409** and (*l*-bornyloxy)aluminum dichloride **411** on the asymmetric induction in the intramolecular Diels–Alder reaction of the triene ester **449** has been reported. Although the cycloadduct was produced in excellent yield under these conditions, it was found to be less than 1% optically pure.[136]

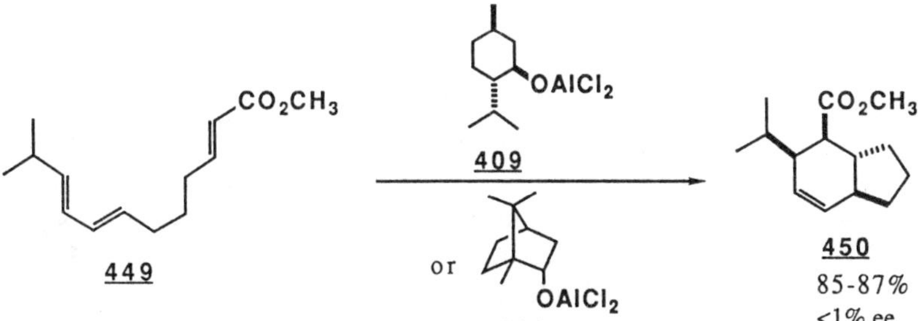

449

or

409

411

450
85-87%
<1% ee

Roush *et al.* have also examined the intramolecular cycloaddition of the triene esters **451a** and **451b** with $(-)$-8-phenylmenthol as the chiral auxiliary in an attempt to induce asymmetry in the overall process. The results of these studies are shown in Table 20. In the case of triene ester **451a**, the 72% de obtained with $TiCl_4$ was more than offset by the 8% isolated yield of cycloadduct. Much better results were obtained with **451b** in combination with (*l*-bornyloxy)aluminum dichloride, which produced **452b** with a 72% de. On the basis of the results obtained with achiral triene ester **449** with the chiral catalyst systems ($<1\%$ de), the chirality of the Lewis acids was assumed to have a negligible effect on the asymmetric induction.[136]

a, R= iPr
b, R= H

Table 20. Lewis Acid Catalyzed Cyclizations of **451**

R	Equivalent of Lewis acid	Temperature °C	% Yield	452 : 453	% de
i-Pr	1.1 $TiCl_4$	23	8	86 : 14	72
i-Pr	0.9 $EtAlCl_2$	23	21	58 : 42	16
i-Pr	1.8 (*d,l*-Menthyloxy)$AlCl_2$	23	75	67 : 33	34
i-Pr	1.6 (*l*-Menthyloxy)$AlCl_2$	23	75	65 : 35	30
i-Pr	1.8 (*l*-Bornyloxy)$AlCl_2$	23	61	67 : 33	34
H	1.8 (*d,l*-Menthyloxy)$AlCl_2$	23	40	75 : 25	50
H	1.5 (*l*-Bornyloxy)$AlCl_2$	23	80	82 : 18	64
H	1.6 (*l*-Bornyloxy)$AlCl_2$	8	72	86 : 14	72

Excellent asymmetric inductions have been realized in intramolecular Diels–Alder reactions related to the one with triene ester **451b**. Both Evans and Oppolzer have utilized the chiral auxiliaries they designed for the intermolecular processes discussed earlier. The reaction of **454a**, with the auxiliary derived from (*S*)-phenylalanol in the presence of Et_2AlCl produced the adducts **455a** and **456a** in a ratio of 95 : 5 (90% de).[35] The corresponding reaction of **454b**, with the camphor sultam auxiliary and $EtAlCl_2$ as the Lewis acid, furnished the adducts **456b** and **455b** in a 97.4:2.5 ratio ($\sim 95\%$ de). A trace amount ($<0.1\%$) of the related exo-adducts was also reported.[137]

Additionally, the reaction of the triene sultam **457** in the presence of EtAlCl$_2$ provided an 81% yield of the decalin derivative **458** in 94.6% de as a 28:1 endo/exo mixture. Simple crystallization furnished adduct **458** in 53% yield with >99% de. The absolute configuration was established by X-ray crystallography.[137]

The Me$_2$AlCl-catalyzed reaction of **459a** provides decalins **460a** and **461a** in 88% yield as a 97:3 diastereomeric mixture. If the opposite sense of induction is desired, either oxazolidinone **459b** or **459c** may be employed. Cycloaddition of **459b** produces a 15:85 mixture of **460b** and **461b** in 70% yield. Oxazolidinone **459c** affords a 3:97 ratio of **460c:461c** in 65% isolated yield.[35b,c]

Oppolzer has also utilized an asymmetric intramolecular Diels–Alder reaction in an enantioselective synthesis of (+)-3-methoxy-1,3,5(10)-estratrien-11,17-dione (465). The benzocyclobutenecarboxylic ester 462 was acylated with the optically active acid chloride 463 to provide the β-keto ester 464. Following decarboxylation and thermolysis at 170°C, a 56% yield of (+)-465 was realized.[138]

An analogous strategy, of trapping o-quinodimethane intermediates, has been employed by Kametani et al. in the asymmetric syntheses of estradiol (468),[139] (−)-3β-hydroxy-17-methoxy-D-homo-18-nor-5α-androsta-13,15, 17-triene (470),[140] and (+)-chenodeoxycholic acid (473).[141]

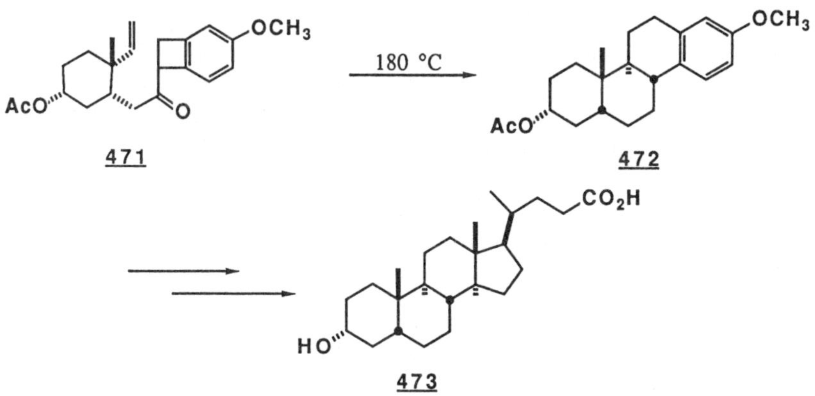

471   180 °C   472

473

The intramolecular trapping of an α,β-unsaturated ketone as a hetero-diene has been successfully utilized in the selective synthesis of (−)-hexahydrocannabinol (**478**).[142] Condensation of (*R*)-citronellal (**474**) with cyclohexanedione **475** in DMF at 100°C afforded adduct **477** in 65% yield as a mixture of epimers at C-3. This was of little consequence as this center is ultimately destroyed in the subsequent aromatization process. The transition state **476**, with the stereocenter derived from citronellal controlling the conformation, has been proposed to account for the observed results. The synthesis of (+)-**478** can be achieved by starting with (*S*)-**474**. Other enantiomeric dihydropyrans have also been prepared by means of this strategy.[143]

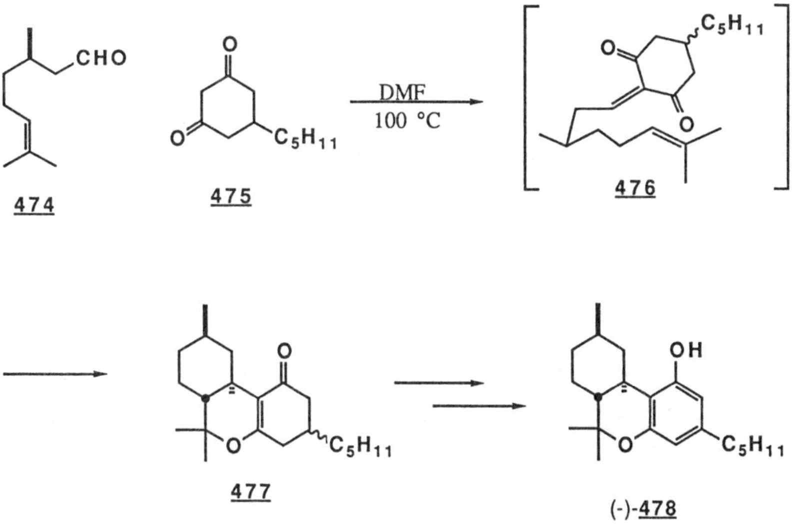

474   475   DMF 100 °C   476

477   (-)-478

Marino and Dax have also synthesized (+)- and (−)-**478** via the intra-molecular cycloaddition of the *o*-quinone methide **480** generated by the 1,4-desilylation–elimination of the bis-silylated *o*-hydroxybenzyl alcohol derivative **479**. This route avoids the necessity of the final aromatization sequence previously employed.[144]

Tietze *et al.*, in their continuing work with the intramolecular hetero-Diels–Alder reaction, have prepared arylideneoxazepane-5,7-dione **482**.[145] They have established via X-ray analysis that the major isomer has the (*Z*)-configuration, contrary to literature reports.[146] Heating **482** with Et$_2$AlCl produces compound **483** in which dienophile adds to the face of the diene *syn* to the bulky substituents.

A synthesis of (+)-farnesiferol C (**488**) by Mukaiyama *et al.* exploited an intriguing cyclization of furan **484**. Under thermal conditions, **484** produces a mixture of **486** and **487** in equal proportions. However, if the alcohol in **484** is first converted to the magnesium alkoxide **485**, the cycloaddition is accelerated and furnishes an 88:12 mixture of the adducts with **486** predominating. The internal chelation not only facilitates the cycloaddition but also imparts a conformational rigidity in the transmission state so the interactions between the α-furyl methylene group and the phenyl substituent are minimized. The adduct **486** was then processed into (+)-**488**.[147]

In a synthetic program aimed at the total synthesis of (−)-cytochalasin C, the intramolecular cycloadditions of chiral (Z)-dienes have been examined. The choice of the (Z)-dienes was made so as to confine the intramolecular Diels–Alder reaction to a single transition stage. The model substrate **489** was assembled from L-(−)-phenylalanine and on thermolysis afforded a 95% yield of **490** as a single isomer. The identity of this adduct was firmly established via an X-ray analysis. The cycloaddition had thus proceeded through conformation **489c** to the exclusion of the alternative conformers **489a** and **489b**.[148]

The Fuchs' group, encouraged by these results, prepared the dienes **491** and **493** and cyclized them in refluxing toluene to provide lactams **492** and **494** in 67 and 69% yield, respectively.[149]

The chiral quinodimethanes generated by the fluoride-induced elimination of the silyl oxazolidinium salts have been intramolecularly trapped to produce an optically active octahydrophenanthrene.[118] The best results were obtained in the reaction of oxazolidinium triflate **495** with CsF at 0°C. Subsequent hydrogenolysis afforded a 71% overall yield of octahydrophenanthrene **496** with a 55% ee.

A chiral (Z)-diene, derived from L-threonine, was used by Ohno and co-workers in the first total synthesis of (+)-actinobolin (**107**). The intramolecular Diels–Alder reaction of diene **497** was effected by heating in benzene at 180°C for 2 h. This rendered adduct **498** in 97% yield. A small amount of an isomer could be detected by 400 MHz ^{1}H-NMR, but the ratio was calculated to be at least 20 to 1 in favor of **498**. This was then converted in a series of steps into (+)-**107**.[150]

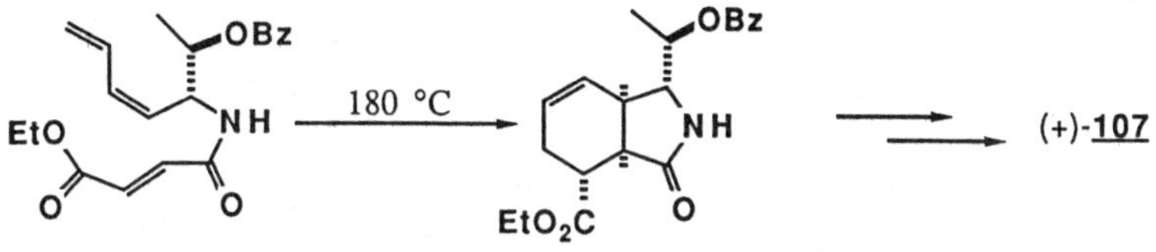

497   180 °C   **498**   (+)-**107**

The decalin system of compactin (**501**) has proven to be an ideal target for the intramolecular Diels–Alder stratagem. Hirama and Uei have successfully completed a chiral total synthesis of (+)-**501**, using trienone **499** in the crucial intramolecular cycloaddition. Cyclization of **499** at 132°C produced the *trans*-decalin **500** in 28% yield along with two cis isomers in 45 and 9% yields. Further transformation of **500** delivered (+)-**501**.[151]

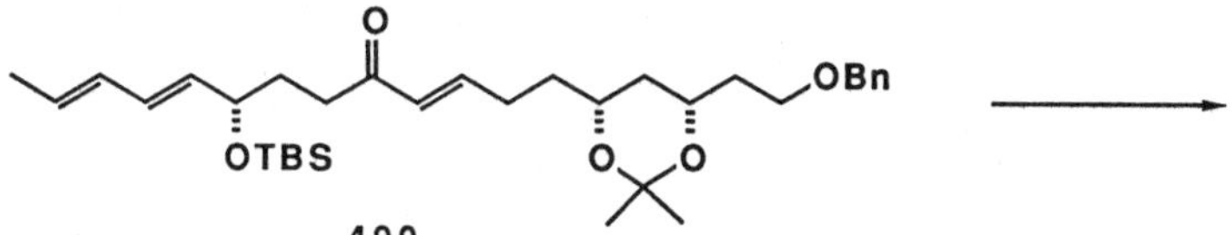

499

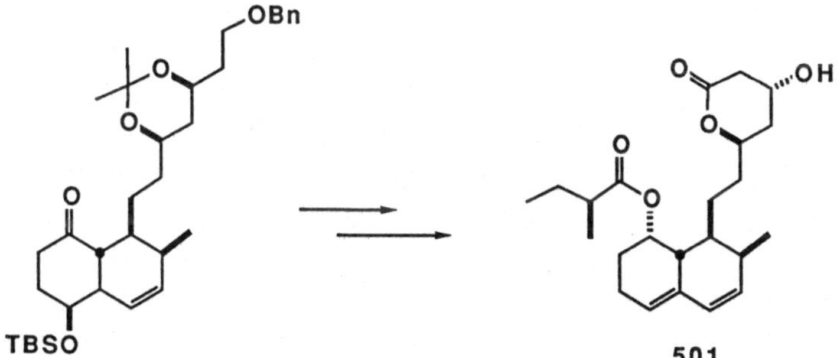

500   **501**

An alternative synthesis of (+)-**501** utilized the cycloaddition of allene **502**. The approach has the advantage that the requisite diene system of the decalin is directly assembled in the cylization. However, it was not surprising the cycloaddition of **502**, with no chiral centers embodied in the tether connecting the diene and dienophile to exert any diastereofacial bias, and subsequent L-selectride reduction produced a mixture of the undesired diastereomer **503** and the desired **504**.[152]

Davidson *et al.* have disclosed an enantiospecific route to the octahydronaphthalene portion of the dihydromevinolin (**507**). The triene **505** was heated at 140°C for 120 h to give the tricyclic lactone **506** as a single isomer in 75% yield.[153]

Three groups have reported syntheses of (−)-indanomycin (**511a**) using an asymmetric intramolecular Diels–Alder reaction to construct the *trans*-fused hexahydroindene skeleton of this novel antibiotic. Nicolaou *et al.* cyclized triene ester **508** by heating it in toluene to afford a 70% yield of compound **509**. This was later coupled with the left wing portion *en route to* completing the synthesis.[154] The other two syntheses introduced the left wing portion prior to cycloaddition. In the Roush synthesis, pentaene **510b** was heated at 60°C to deliver in 51% yield the methyl ester **511b**.[155] The Boeckman synthesis thermolyzed pentaene **510c** at 65°C to produce adduct **511c** in 53% yield. This was transformed in two steps to (−)-**511a**.[156]

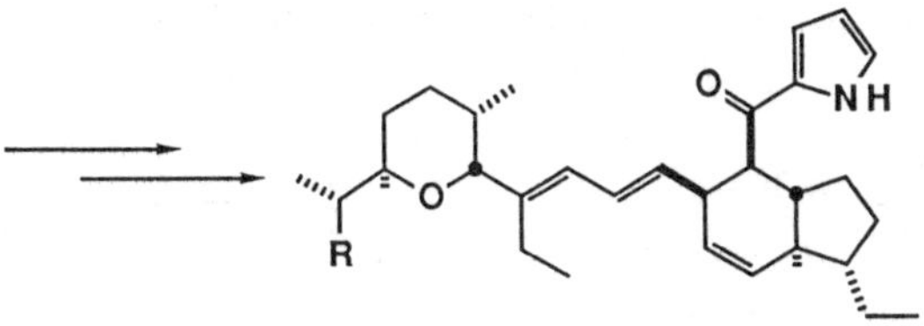

511

a, R= CO₂H

b, R= CO₂CH₃

c, R= CH₂OMOM

510b, c → Δ → **511b, c**

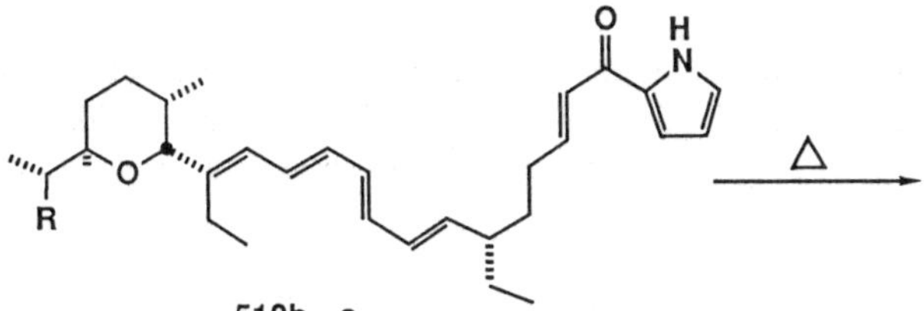

Kitahara *et al.* have communicated enantioselective syntheses of sclero-porin (**514a**) and sclerosporal (**514b**) exploiting an intramolecular cyclo-ddition. The *cis*-octalone **513** was prepared by the cyclization of trienone 12, which was derived from (−)-carvone. The octalone **513** was trans-ormed in a straightforward manner to **514a** and **514b**.[157]

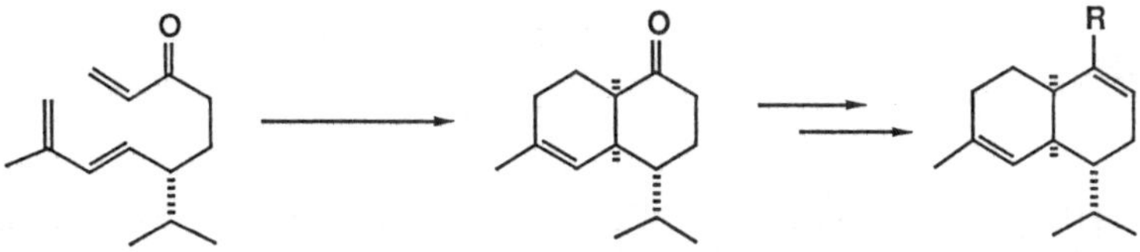

512   **513**   **514**   a, R= CO₂H

b, R= CHO

The total synthesis of (−)-α-selinene (**517**) from the (−)-carvone derived ieneone **515** has been reported by Caine and Stanhope. Cycloaddition of 5 using thermal conditions produced a 15:7:3 mixture of adducts with the ιns-fused octalin **516** being the major component. The other two products ere the two *cis*-fused octalins. The Wolff–Kishner reduction of **516** yielded ιtural product **517**.[158]

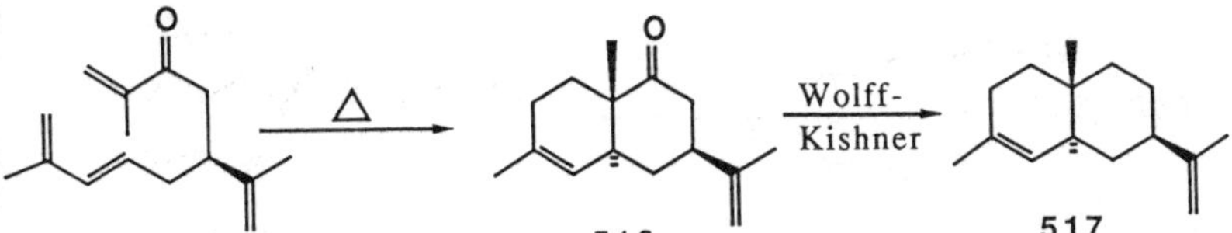

515   **516**   **517**

Wolff-Kishner

Two independent syntheses of the two optical antipodes of the active component in catnip oil, (+)- and (−)-nepetalactone (520), utilizing the intramolecular [4 + 2] cycloaddition of α,β-unsaturated aldehydes as hetero dienes have appeared. In Denmark's synthesis, the (Z)-heterodiene 518 under the influence of BF₃·Et₂O, was trapped intramolecularly by the ketene thioacetal to provide adduct 519 in 55% yield as a single diastereomer Hydrolysis with HgO gave (+)-520 in 76% yield.[159] When Schreiber treated the (E)-enal aldehyde 521 with N-methylaniline, the dihydropyran 522 was obtained, via cycloaddition of the intermediate enamine, in 84% yield as a 10:1 mixture of methyl epimers. Hydrolysis and oxidation of 522 gave rise to (−)-520.[160]

Thomas and Whitehead have synthesized cytochalasin H (527) via a intramolecular Diels–Alder reaction to close the 11-membered ring. The required substrate for the cycloaddition was assembled by coupling the triene-imidizole 523 with pyrolidinone 524. A selenation–oxidation–elimination sequence installed the requisite unsaturation into dienophile 525 Thermolysis of 525 in toluene at 80–100°C for 5 h afforded the Diels–Alder adduct 526 as a single isomer in 38% yield.[161a] Cytochalasin H (527) was secured from 526 following a sequence of functional group manipulations.[161b]

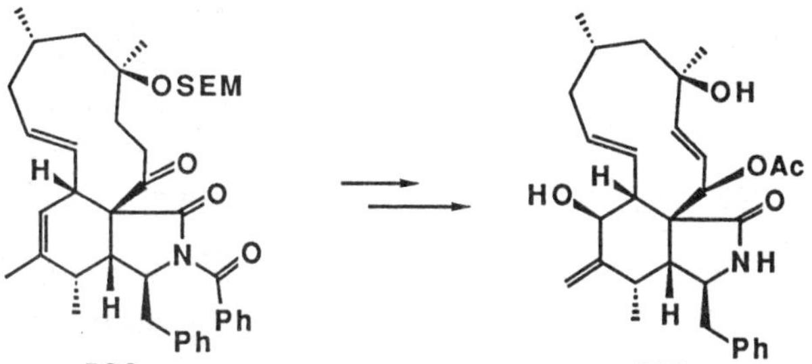

526 → **527**

The first reported synthesis of (+)-7,10-diisocyanoadociane (**533**) in which two intramolecular Diels–Alder reactions were utilized has been communicated by Corey and Magriotis. The initial stereocenters were established in **528** via a diastereoselective Michael addition. This compound was converted to triene **529** and cyclized to the *trans*-fused decalin **530** at 150°C. Processing of **530** to triene **531** and thermolysis at 185°C afforded the desired perhydropyrene **532** in 54% yield along with 36% of a diastereomeric adduct. This was transformed in a series of reactions to (+)-**533**.[162]

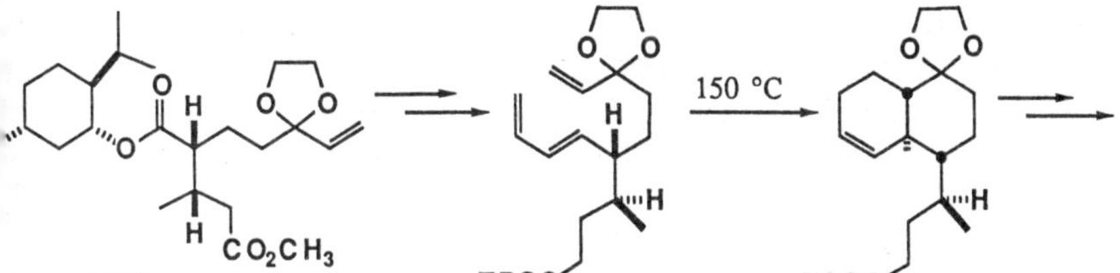

528 **529** **530**

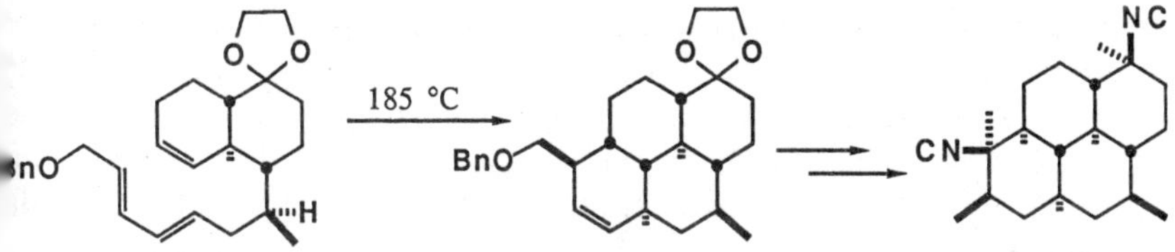

531 → **532** → **533**

In an approach directed toward the synthesis of nargenicin A_1 (**538**), Roush and Coe examined the intramolecular [4 + 2] cycloaddition of the chiral triene ester **534** to assemble the requisite *cis*-fused decalin nucleus. The cycloaddition reaction produced the *cis*-fused decalin **535** as the only detectable product. Unfortunately, this was not the diastereomer required for future elaborations toward **538**. The reaction had therefore proceeded through the boat-like transition state **537**, rather than the expected chair-like transition state **536**.[163]

Two independently conducted synthetic studies by Marshall and Roush aimed at the macrocyclic antitumor antibiotics chlorothricolide (**539**), kijanolide (**540**), and tetronolide (**541**) utilize intramolecular Diels–Alder reactions to assemble the octahydronaphthalene portions of these natural products. In Marshall's approach the trieneals **543** and **546** were cyclized under Lewis acid conditions. The (*R*)-methyl-3-hydroxy-2-methyl-propanoate (**542**)-derived trienal **543** when treated with EtAlCl$_2$ cyclizes to furnish a 1:1 mixture of adducts **544** and **545** that are epimeric at C-8. These adducts were shown to have enantiomeric excesses of 88 and 82%, respectively.[164a] Trienal **546** cyclizes under thermal or Lewis acid catalysis to afford octahydronaphthalene **547** as a 90:10 mixture of C-8 epimers.[164b]

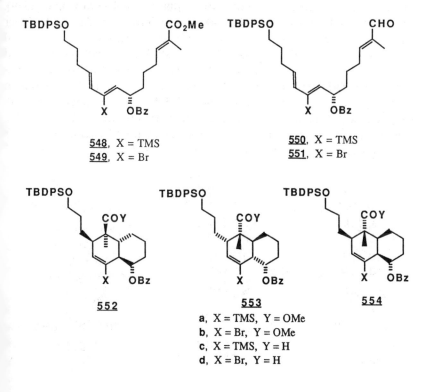

Roush and Riva studied the cycloaddition of trienes 548–551, which produces mixtures of cycloadducts **552–554**. The results of this study are shown in Table 21. An evaluation of the triene preparations, yield of cycloadduct, and ease of functional group manipulation in the cycloadduct revealed that bromo-triene **549** was the optimal precursor.[165]

548, X = TMS
549, X = Br

550, X = TMS
551, X = Br

552

553

554

a, X = TMS, Y = OMe
b, X = Br, Y = OMe
c, X = TMS, Y = H
d, X = Br, Y = H

Table 21. Cycloadditions of **548–551**

Triene	Conditions	552 : 553 : 554	Combined yield, %
548	160°C	78 : 8 : 14	85
549	160°C	62 : 4 : 34	80
550	160°C	79 : 9 : 12	82
550	Et$_2$AlCl, −15°C	89 : 5 : 6	77
550	EtAlCl$_2$, −40°C	90 : 5 : 5	32
551	160°C	75 : 3 : 22	86
551	EtAlCl$_2$, −15°C	90 : 1 : 9	24

A French group has developed two related approaches to the *Strychnos* alkaloids. In the first approach, the L-tryptophan-derived diester **555** was refluxed in toluene in the presence of butyraldehyde and a large excess of acetic acid to produce a 68% yield of the two β-anilinoacrylates **556** and **557** in a ratio of 85:15. The structure of the major product **556**, which corresponds to the (+)-series, was verified by X-ray crystallography.[166a] The (−)-series is potentially available using this strategy but relative unavailability D-tryptophan makes it less attractive. An alternative was devised starting from the chiral tryptamine **558**. Processing **558** through the same sequence as for the other series affords a 50% yield of β-anilinoacrylate **559**.[166b]

Stork and Saccomano developed an enantioselective route to adenosterone (**564**) utilizing the internal Diels–Alder reaction. An intramolecular Michael addition of **560** using the auxiliary **273** induced the desired chirality in **561**. The diester **561** was transformed into trienone **562** and cylized to furnish tetracycle **563**. This was converted as previously described into **564**.[167]

A D-xylose-derived triene **565** was thermolyzed as a 4:6 mixture of $E:Z$ dienes to produce a single adduct in 83% yield. This remarkable selectivity is possibly due to thermal isomerization for the dienes prior to cycloaddition. Cycloadduct **566** is potentially useful for the preparation of prostaglandins, enzyme inhibitors, and pseudo sugars.[168]

A synthesis of (+)-sterpurene (**569**) utilizing the intramolecular transfer of central to axial to central chiral elements has been reported by Gibbs and Okamura. The optically active propargylic alcohol **567** was reacted with PhSCl to furnish the desired cycloadduct **568** in 70% yield as a 61:39 mixture of sulfoxide diastereomers. This was converted in a short series of steps into (+)-**569**.[169]

VII. CONCLUSION

The asymmetric Diels–Alder reaction has come a long way since the initial investigations in terms of the control of absolute stereochemistry. The original cycloadditions, which only produced meager inductions, have given way to processes capable of attaining essentially complete asymmetric induction. This is not meant to imply the state of the art has advanced to a level at which there is no need for further developments. There is always room for improvements. This chapter is intended to leave the reader with an appreciation of the power and versatility of the asymmetric Diels–Alder reaction and at the same time lay a foundation on which new and potentially more exciting ideas can be appended.

ACKNOWLEDGMENTS

I would like to thank my graduate students Anand Aminbhavi, D. Jeff Black, Zacharia Cheruvallath, Patrick Cyr, Jeff Smith, and Nancy Rach for their able-bodied assistance in compiling the references needed to write this chapter. I would also thank Dr. Josie Reed for her expertise in the editorial proofreading of this chapter. Lastly, I would like to extend my gratitude to the editor and good friend, T. Hudlicky.

REFERENCES

1. Diels, O.; Alder, K. *Justus Liebigs Ann. Chem.* **1928**, *460*, 98.
2. Overman, L. E.; Taylor, G. F.; Houk, K. N.; Domelsmith, L. N. *J. Am. Chem. Soc.* **1978**, *100*, 3182, and references cited therein.
3. (a) Martin, J. G.; Hill, R. K. *Chem. Rev.* **1961**, *61*, 537. (b) Kokube, Y.; Furukawa, J.; Fueno, T. *J. Am. Chem. Soc.* **1970**, *92*, 6548. (c) *Ibid.* **1972**, *94*, 3633. (d) Sugimoto, T.; Kokube, Y.; Furukawa, J. *Tetrahedron Lett.* **1976**, 1587. (e) Lee, M. W.; Herndon, W. C. *J. Org. Chem.* **1978**, *43*, 518.
4. For some exceptions see: Firestone, R. A. *Tetrahedron Lett.* **1977**, *23*, 3009.

5. Morrison, J. D. *Asymmetric Synthesis*, Academic Press, New York: (a) Vol. 1, 1983; (b) Vol. 2, 1983; (c) Vol. 3, 1984. (d) Vol. 4, 1984.

6. (a) Onischenko, A. S. *Diene Synthesis*, Israel Program of Scientific Translations, Daniel Davy & Co., New York, 1964. (b) Wollweber, H. *Diels–Alder Reaction*, Georg Thieme Verlag, Stuttgart, 1972. (c) Sauer, J. *Angew. Chem. Int. Ed. Engl.* **1966**, *5*, 211. (d) *Ibid.* **1967**, *6*, 16. (e) Sauer, J.; Sustmann, R. *Angew. Chem. Int. Ed. Engl.* **1980**, *19*, 779. (f) Hamer, J. *1,4-Cycloaddition Reactions*, Academic Press, New York, 1967. (g) Titov, Y. A. *Russ. Chem. Rev.* **1962**, *31*, 267. (h) Seltzer, S. In *Advances in Alicylcic Chemistry*, Hart, H.; Karabatsos, G. J., eds., Academic Press, New York, 1968, Vol. 2. (i) Petrzilka, M.; Grayson, J. I. *Synthesis* **1981**, 753.

7. (a) Carlson, R. G. *Ann. Rep. Med. Chem.* **1974**, *9*, 270. (b) Mehta, G. *J. Chem. Ed.* **1976**, *53*, 551. (c) Oppolzer, W. *Angew. Chem. Int. Ed. Engl.* **1977**, *16*, 10. (d) Brieger, G.; Bennett, J. N. *Chem. Rev.* **1980**, *80*, 63. (e) Taber, D. F. *Intramolecular Diels–Alder and Aler Ene Reactions*, Springer-Verlag, Berlin, 1984. (f) Ciganek, E. *Org. React.* **1984**, *32*, 1.

8. (a) Oppolzer, W. *Angew. Chem. Int. Ed. Engl.* **1984**, *23*, 876. (b) Paquette, L. A. In *Asymmetric Synthesis*, Morrison, J. D., ed., Academic Press, Orlando, FL, 1984, Vol. 3B, pp. 455–501. (c) Helmchen, G.; Karge, R.; Weetman, J. In *Modern Synthetic Methods*, Scheffold, R., ed.; Springer-Verlag, Berlin, Heidelberg, 1986, Vol. 4. pp. 262–306.

9. The first experiments were actually conducted by Korolev and Mur, but were found to be erroneous by Walborsky, see: Korolev, A.; Mur, V. *Dokl. Akad. Nauk S.S.S.R.* **1949**, *59*, 251.

10. Walborsky, H. M.; Barash, L.; Davis, T. C. *J. Org. Chem.* **1961**, *26*, 4778.

11. Walborsky, H. M.; Barash, L.; Davis, T. C. *Tetrahedron* **1963**, *19*, 2333.

12. Jurczak, J. *Bull. Chem. Soc. Jpn.* **1979**, *52*, 3438.

13. Helmchen, G.; Schmierer, R. *Angew. Chem. Int. Ed. Engl.* **1981**, *20*, 205.

14. (a) Tolbert, L. M.; Ali, M. B. *J. Am. Chem. Soc.* **1981**, *103*, 2104. (b) *Ibid.* **1984**, *106*, 3810.

15. (a) Sauer, J.; Kredel, J. *Tetrahedron Lett.* **1966**, 6359. (b) Farmer, R. F.; Hamer, J. *J. Org. Chem.* **1966**, *31*, 2418.

16. Cervinka, O.; Kriz, O. *Coll. Czech. Chem. Commun.* **1968**, *33*, 2342.

17. Corey, E. J.; Ensley, H. E. *J. Am. Chem. Soc.* **1975**, *97*, 6908.

18. Oppolzer, W.; Kurth, M.; Reichlin, D.; Moffatt, F. *Tetrahedron Lett.* **1981**, *22*, 2545.

19. Le Drain, C.; Greene, A. E. *J. Am. Chem. Soc.* **1982**, *104*, 5473.

20. Boeckman, R. K.; Naegley, P. C.; Arthur, S. D. *J. Org. Chem.* **1980**, *45*, 754.

21. Oppolzer, W.; Kurth, M.; Reichlin, D.; Chapuis, C.; Mohnhaupt, M.; Moffatt, F. *Helv. Chim. Acta* **1981**, *64*, 2802.

22. Oppolzer, W.; Chapuis, C.; Dao, G. M.; Reichlin, D.; Godel, T. *Tetrahedron Lett.*, **1982**, *23*, 4781.

23. Oppolzer, W. In W. Bartmann, B. M. Trost: *Selectivity—a Goal for Synthetic Efficiency*, Workshop Conferences Hoechst, Vol. 14, Verlag Chemie, Weinheim, 1984, pp. 137–167.

24. Loncharich, R. J.; Schwartz, T. R.; Houk, K. N. *J. Am. Chem. Soc.* **1987**, *109*, 14.

25. Oppolzer, W.; Chapuis, C. *Tetrahedron Lett.* **1983**, *24*, 4665.

26. *Ibid.* **1984**, *25*, 5383.

27. Oppolzer, W.; Chapuis, C.; Kelly, M. J. *Helv. Chim. Acta* **1983**, *66*, 2358.

28. Oppolzer, W.; Chapuis, C.; Bernardinelli, G. *Tetrahedron Lett.* **1984**, *25*, 5885.

29. Curran, D. P.; Kim, B. H.; Piyasena, H. P.; Loncharich, R. J.; Houk, K. N. *J. Org. Chem.* **1987**, *52*, 2137.

30. Oppolzer, W.; Kelly, M. J.; Bernardinelli, G. *Tetrahedron Lett.* **1984**, *25*, 5889.

31. Oppolzer, W.; Chapuis, C.; Bernardinelli, G. *Helv. Chim. Acta* **1984**, *67*, 1397.

32. Vandewalle, M.; Van der Eyken, J.; Oppolzer, W.; Vullioud, C. *Tetrahedron* **1986**, *42*, 4035.

33. Choy, W.; Reed, L. A.; Masamune, S. *J. Org. Chem.* **1983**, *48*, 1139.

34. Masamune, S.; Reed, L. A.; Davis, J. T.; Choy, W. *J. Org. Chem.* **1983**, 48, 4441.
35. (a) Evans, D. A.; Chapman, K. T.; Bisaha, J. *J. Am. Chem. Soc.* **1984**, *106*, 4261. (b) Evans, D. A.; Chapman, K. T.; Bisaha, J. *Tetrahedron Lett.* **1984**, *25*, 4071. (c) Evans, D. A.; Chapman, K. T.; Bisaha, J. *J. Am. Chem. Soc.* **1988**, *110*, 1238.
36. Schmierer, R. Ph.D. Dissertation, University of Stuttgart, 1980.
37. (a) Poll, T.; Helmchen, G.; Bauer, B. *Tetrahedron Lett.* **1984**, *25*, 2191. (b) Helmchen, G.; Ihrig, K.; Schindler, H. *Tetrahedron Lett.* **1987**, *28*, 183.
38. Poll, T.; Metter, J. O.; Helmchen, G. *Angew. Chem. Int. Ed. Engl.* **1985**, *24*, 112.
39. Poll, T.; Sobczak, A.; Hartmann, H.; Helmchen, G. *Tetrahedron Lett.* **1985**, *26*, 3095.
40. Shing, T. K. M.; Lloyd-Williams, P. *J. C. S. Chem. Commun.* **1987**, 423.
41. Jones, G. *Tetrahedron Lett.* **1974**, 2231.
42. (a) Mattes, K. C.; Hsia, M. T.; Hutchinson, C. R.; Sisk, S. A. *Tetrahedron Lett.* **1977**, 3541. (b) Hutchinson, C. R.; Sisk, S. A. *J. Org. Chem.* **1979**, *44*, 3500.
43. (a) Jurczak, J. *Pol. J. Chem.* **1979**, *53*, 209. (b) Jurczak, J.; Tkacz, M. *Synthesis* **1979**, 42.
44. Primeau, J. L.; Anderson, R. C.; Fraser-Reid, B. *J. C. S. Chem. Commun.* **1980**, 6.
45. Rahman, M. A.; Fraser-Reid, B. *J. Am. Chem. Soc.* **1985**, *107*, 5576.
46. For other examples see: Fraser-Reid, B.; Underwood, R.; Osterhout, M.; Grossman, J. A.; Liotta, D. *J. Org. Chem.* **1986**, *51*, 2152, and references cited therein.
47. Isobe, M.; Nishikawa, T.; Pikul, S.; Goto, T. *Tetrahedron Lett.* **1987**, *28*, 6485.
48. Mann, J.; Thomas, A. *J. C. S. Chem. Commun.* **1985**, 737.
49. (a) Ortuno, R. M.; Corbera, J.; Font, J. *Tetrahedron Lett.* **1986**, *27*, 1081. (b) Ortuno, R. M.; Batllori, R.; Ballesteros, M.; Monsalvatje, M.; Corbera, J.; Sanchez-Ferrando, F.; Font, J. *Tetrahedron Lett.* **1987**, *28*, 3405. (c) Ortuno, R. M.; Batllori, R.; Ballesteros, M.; Corbera, J.; Sanchez-Ferrando, F.; Font, J. *Tetrahedron* **1988**, *44*, 1711. (d) Takano, S.; Inomata, K.; Kurotaki, A.; Ohkawa, T.; Ogasawara, K. *J. C. S. Chem. Commun.* **1987**, 1720.
50. Franck, R. W.; John, T. V. *J. Org. Chem.* **1980**, *45*, 1172.
51. Franck, R. W.; John, T. V.; Olejniczak, K.; Blount, J. F. *J. Am. Chem. Soc.* **1982**, *104*, 1106.
52. (a) Horton, D.; Machinami, T. *J. C. S. Chem. Commun.* **1981**, 88. (b) Horton, D.; Machinami, T.; Takagi, Y. *Carbohydrate Res.* **1983**, *121*, 135.
53. Mulzer, J.; Kappert, M.; Huttner, G.; Jibril, I. *Tetrahedron Lett.* **1985**, *26*, 1631.
54. Takano, S.; Kurotaki, A.; Ogasawara, K. *Tetrahedron Lett.* **1987**, *28*, 3991.
55. Sundin, A.; Frejd, T.; Magnusson, G. *Tetrahedron Lett.* **1985**, *26*, 5605.
56. Fitzsimmons, B. J.; Leblanc, Y.; Rokach, J. *J. Am. Chem. Soc.* **1987**, *109*, 285.
57. Nerdel, F.; Dahl, H. *Justus Liebigs Ann. Chem.* **1967**, *710*, 90.
58. Harayama, T.; Cho, H.; Inubushi, Y. *Tetrahedron Lett.* **1975**, 2693.
59. Angell, E. C.; Fringuelli, F.; Pizzo, F.; Porter, B.; Taticchi, A.; Wenkert, E. *J. Org. Chem.* **1985**, *50*, 4696.
60. The uncatalyzed reaction gave only a 7% yield of regioisomers of undefined stereochemistry, see Ref. 52.
61. Smith, A. B.; Liverton, N. J.; Hrib, N. J.; Sivaramakrishnan, H.; Winzenberg, K. *J. Am. Chem. Soc.* **1986**, *108*, 3040.
62. Feringa, B. L.; de Jong, J. C. *J. Org. Chem.* **1988**, *53*, 1125.
63. Yamauchi, M.; Watanabe, T. *J. C. S. Chem. Commun.* **1988**, 27.
64. Katagari, N.; Haneda, T.; Hayasaka, E.; Watanabe, N.; Kaneko, C. *J. Org. Chem.* **1988**, *53*, 227.
65. Maignan, C.; Raphael, R. A. *Tetrahedron* **1983**, *39*, 3245.
66. Maignan, C.; Guessous, A.; Rouessac, F. *Tetrahedron Lett.* **1984**, *25*, 1727.
67. Koizumi, T.; Hakamada, I.; Yoshi, E. *Tetrahedron Lett.* **1984**, *25*, 87.
68. Arai, Y.; Kuwayama, S.; Takeuchi, Y.; Koizumi, T. *Tetrahedron Lett.* **1985**, *26*, 6205.

69. Arai, Y.; Kuwayama, S.; Takeuchi, Y.; Koizumi, T. *Synthetic Commun.* **1986**, *16*, 233.
70. Maignan, C.; Belkasmioui, F. *Tetrahedron Lett.* **1988**, *29*, 2823.
71. Takayama, H.; Iyobe, A.; Koizumi, T. *J. C. S. Chem. Commun.* **1986**, 771.
72. Takayama, H.; Hayashi, K.; Koizumi, T. *Tetrahedron Lett.* **1986**, *27*, 5509.
73. De Lucchi, O.; Marchioro, C.; Valle, G.; Modena, G. *J. C. S. Chem. Commun.* **1985**, 878.
74. De Lucchi, O.; Lucchini, V.; Marchioro, C.; Valle, G.; Modena, G. *J. Org. Chem.* **1986**, *51*, 1457.
75. Kahn, S. D.; Hehre, W. J. *Tetrahedron Lett.* **1986**, *27*, 6041.
76. Corey, E. J.; Shibasaki, M.; Nicolaou, K. C.; Malmsten, C. L.; Samuelsson, B. *Tetrahedron Lett.* **1976**, 737.
77. Posner, G. H.; Wettlaufer, D. G. *Tetrahedron Lett.* **1986**, *27*, 667.
78. Gruseck, U.; Heuschmann, M. *Tetrahedron Lett.* **1987**, *28*, 2681.
79. Bodalski, R.; Koszuk, J.; Krawczyk, H.; Pietrusiewicz, K. M. *J. Org. Chem.* **1982**, *47*, 2219.
80. Davies, S. G.; Walker, J. C. *J. C. S. Chem. Commun.* **1986**, 609.
81. Jurczak, J.; Baranowski, B. *Pol. J. Chem.* **1978**, *52*, 1857.
82. Jurczak, J. *Pol. J. Chem.* **1979**, *53*, 2539.
83. Jurczak, J.; Tkacz, M. *J. Org. Chem.* **1979**, *44*, 3347.
84. Danishefsky, S.; Kobayashi, S.; Kerwin, J. F. *J. Org. Chem.* **1982**, *47*, 1983.
85. Danishefsky, S.; Chao, K.-H.; Schulte, G. *J. Org. Chem.* **1985**, *50*, 4650.
86. Danishefsky, S.; Barbachyn, M. *J. Am. Chem. Soc.* **1985**, *107*, 7761.
87. (a) Danishefsky, S.; Selnick, H. G.; DeNinno, M. P.; Zelle, R. E. *J. Am. Chem. Soc.* **1987**, *109*, 1572 (b) Danishefsky, S.; Selnick, H. G.; DeNinno, M. P.; Zelle, R. E. *J. Am. Chem. Soc.* **1988**, *110*, 4368.
88. (a) Danishefsky, S.; Armistead, D. M.; Wincott, F. E.; Selnick, H. G.; Hungate, R. *J. Am. Chem. Soc.* **1987**, *109*, 8117. (b) Danishefsky, S.; Selnick, H. G.; Armistead, D. M.; Wincott, F. E. *J. Am. Chem. Soc.* **1987**, *109*, 8119.
89. (a) Garner, P. *Tetrahedron Lett.* **1984**, *25*, 5855. (b) Garner, P.; Ramakanth, S. *J. Org. Chem.* **1986**, *51*, 2609.
90. Golebiowski, A.; Jacobsson, U.; Jurczak, J. *Tetrahedron* **1987**, *43*, 3063.
91. Nitsch, H.; Kresze, G. *Angew. Chem. Int. Ed. Engl.* **1976**, *15*, 760.
92. Sabuni, M.; Kresze, G.; Braun, H. *Tetrahedron Lett.* **1984**, *25*, 5377.
93. (a) Gardlik, J. M.; Paquette, L. A. *Tetrahedron Lett.* **1979**, 3597. (b) Horn, K. A.; Browne, A. R.; Paquette, L. A. *J. Org. Chem.* **1980**, *45*, 5381.
94. Paquette, L. A.; Doehner, R. F. *J. Org. Chem.* **1980**, *45*, 5105.
95. Gardlik, J. M.; Paquette, L. A. *J. Am. Chem. Soc.* **1980**, *102*, 5016.
96. (a) Paquette, L. A.; Gardlik, J. M.; Johnson, L. K.; McCullough, K. J. *J. Am. Chem. Soc.* **1980**, *102*, 5026. (b) Paquette, L. A.; Hanzawa, Y.; McCullough, K. J.; Tagle, B.; Swenson, W.; Clardy, J. *J. Am. Chem. Soc.* **1981**, *103*, 2262. (c) Paquette, L. A.; Hanzawa, Y.; Hefferon, G. J.; Blount, J. F. *J. Org. Chem.* **1982**, *47*, 265.
97. Klobucar, W. D.; Paquette, L. A.; Blount, J. F. *J. Org. Chem.* **1981**, *46*, 4021.
98. Paquette, L. A.; Doehner, R. E.; Jenkins, J. A.; Blount, J. F. *J. Am. Chem. Soc.* **1980**, *102*, 1188.
99. Jenkins, J. A.; Doehner, R. E.; Paquette, L. A. *J. Am. Chem. Soc.* **1980**, *102*, 2131.
100. Porskamp, P. A. T. W.; Haltiwanger, R. C.; Zwanenburg, B. *Tetrahedron Lett.* **1983**, *24*, 2035.
101. Van den Broek, L. A. G. M.; Porskamp, P. A. T. W.; Haltiwanger, R. C.; Zwanenburg, B. *J. Org. Chem.* **1984**, *49*, 1691.
102. Whitesell, J. K.; James, D.; Carpenter, J. F. *J. C. S. Chem. Commun.* **1985**, 1449.
103. Remiszewski, S. W.; Yang, J.; Weinreb, S. M. *Tetrahedron Lett.* **1986**, *27*, 1853.
104. Larsen, S. D.; Grieco, P. A. *J. Am. Chem. Soc.* **1985**, *107*, 1768.

105. David, S.; Eustache, J.; Lubineau, A. *J. Chem. Soc. Perkin I* **1974**, 2274.
106. David, S.; Lubineau, A.; Thieffry, A. *Tetrahedron* **1978**, *34*, 299.
107. David, S. Lubineau, A.; Vatele, J. -M. *J. C. S. Chem. Commun.* **1978**, 535.
108. David, S.; Lubineau, A. *Nouv. J. Chem.* **1977**, *1*, 375.
109. David, S.; Eustache, J.; Lubineau, A. *J. Chem. Soc. Perkin I* **1979**, 1795.
110. David, S.; Eustache, J. *J. Chem. Soc. Perkin I* **1979**, 2521.
111. (a) Gupta, R. C.; Harland, P. A.; Stoodley, R. J. *J. C. S. Chem. Commun.* **1983**, 754. (b) Gupta, R. C.; Harland, P. A.; Stoodley, R. J. *Tetrahedron* **1984**, *40*, 4657. (c) Gupta, R. C.; Slawin, A. M. Z.; Stoodley, R. J.; Williams, D. J. *J.C.S.Chem. Commun.* **1986**, 668.
112. Trost, B. M.; Godleski, S. A.; Genet, J. P. *J. Am. Chem. Soc.* **1978**, *100*, 3930.
113. Trost, B. M.; O'Krongly, D.; Belletire, J. L. *J. Am. Chem. Soc.* **1980**, *102*, 7595
114. Cohen, N.; Banner, B. L.; Eichel, W. F.; Valenta, Z.; Dickinson, R. A. *Synthetic Commun.* **1978**, *8*, 427.
115. Dauben, W. G.; Bunce, R. A. *Tetrahedron Lett.* **1982**, *23*, 4875.
116. Tietze, L.-F.; Glusenkamp, K.-H. *Angew. Chem. Int. Ed. Engl.* **1983**, *22*, 887.
117. Schmidt, R. R.; Frick, W.; Haag-Zeino, B.; Apparao, S. *Tetrahedron Lett.* **1987**, *28*, 4045.
118. Ito, Y.; Amino, Y.; Nakatsuka, M.; Saegusa, T. *J. Am. Chem. Soc.* **1983**, *105*, 1586.
119. (a) Charlton, J. L. *Tetrahedron Lett.* **1985**, *26*, 3413. (b) Charlton, J. L. *Can. J. Chem.* **1986**, *64*, 720. (c) Charlton, J. L.; Alauddin, M. M. *Tetrahedron* **1987**, *43*, 2873.
120. Schomburg, D.; Theilmann, M.; Winterfeldt, E. *Tetrahedron Lett.* **1985**, *26*, 1705.
121. Schomburg, D.; Theilmann, M.; Winterfeldt, E. *Tetrahedron Lett.* **1986**, *27*, 5833.
122. (a) Kozikowski, A. P.; Nieduzak, T. R.; Springer, J. P. *Tetrahedron Lett.* **1986**, 27, 819. (b) Kozikowski, A. P.; Konoike, T.; Nieduzak, T. R. *J. C. S. Chem. Commun.* **1986**, 1350. (c) Kozikowski, A. P.; Nieduzak, T. R.; Konoike, T.; Springer, J. P. *J. Am. Chem. Soc.* **1987**, *109*, 5167.
123. Brandes, E.; Grieco, P. A.; Garner, P. *J. C. S. Chem. Commun.* **1988**, 500.
124. (a) Bednarski, M.; Danishefsky, S. *J. Am. Chem. Soc.* **1983**, *105*, 3716. (b) Bednarski, M.; Danishefsky, S. *J. Am. Chem. Soc.* **1983**, *105*, 6968. (c) Bednarski, M.; Danishefsky, S. *J. Am. Chem. Soc.* **1986**, *108*, 7060.
125. Bednarski, M.; Maring, C.; Danishefsky, S. *Tetrahedron Lett.* **1983**, *24*, 3451.
126. Guseinov, M. M.; Akmedov, I. M.; Mamedov, E. C. *Azerb. Khim. Zhur.* **1976**, 46.
127. Hashimoto, S.; Komeshima, N.; Koga, K. *J. C. S. Chem. Commun.* **1979**, 437.
128. Kobayashi, E.; Matsumura, S.; Furukawa, J. *Polym. Bull.* **1980**, *3*, 285.
129. Quimpere, M.; Jankowski, K. *J. C. S. Chem. Commun.* **1987**, 676.
130. Kaufmann, D.; Bir, G. *Tetrahedron Lett.* **1987**, *28*, 777.
131. (a) Narasaka, K.; Inoue, M.; Okada, N. *Chem. Lett.* **1986**, 1109. (b) Narasaka, K.; Inoue, M.; Okada, N. *Chem. Lett.* **1986**, 1967.
132. Chapuis, C.; Jurczak, J. *Helv. Chim. Acta* **1987**, *70*, 436.
133. Kelly, T. R.; Whiting, A.; Chandrakumar, N. S. *J. Am. Chem. Soc.* **1986**, *108*, 3510.
134. Maruoka, K.; Sakurai, M.; Fujiwara, J.; Yamamoto, H. *Tetrahedron Lett.* **1986**, *27*, 4895.
135. Marouka, K.; Itoh, T.; Shirasaka, T.; Yamamoto, H. *J. Am. Chem. Soc.* **1988**, *110*, 310.
136. Roush, W. R.; Gillis, H. R.; Ko, A. I. *J. Am. Chem. Soc.* **1982**, *104*, 2269.
137. Oppolzer, W.; Dupuis, D. *Tetrahedron Lett.* **1985**, *26*, 5437.
138. Oppolzer, W.; Battig, K.; Petrzilka, M. *Helv. Chim. Acta* **1978**, *61*, 1945.
139. (a) Kametani, T.; Matsumoto, H.; Nemoto, H.; Fukumoto, K. *Tetrahedron Lett.* **1978**, 2425. (b) Kametani, T.; Matsumoto, H.; Nemoto, H.; Fukumoto, K. *J. Am. Chem. Soc.* **1978**, *100*, 6218.
140. Kametani, T.; Suzuki, K.; Nemoto, H. *J. Org. Chem.* **1980**, *45*, 2204.
141. Kametani, T.; Suzuki, K.; Nemoto, H. *J. Org. Chem.* **1982**, *47*, 2331.

142. (a) Tietze, L.-F.; v. Kiedrowski, G.; Harms, K.; Clegg, W.; Sheldrick, G. *Angew. Chem. Int. Ed. Engl.* **1980**, *19*, 134. (b) Tietze, L.-F.; v. Kiedrowski, G.; Berger, B. *Angew. Chem. Int. Ed. Engl.* **1982**, *21*, 221.

143. Tietze, L.-F.; v. Kiedrowski, G. *Tetrahedron Lett.* **1981**, 219.

144. Marino, J. P.; Dax, S. L. *J. Org. Chem.* **1984**, *49*, 3672.

145. (a) Tietze, L. F.; Brand, S.; Pfeiffer, T. *Angew. Chem. Int. Ed. Engl.* **1985**, 24, 784. (b) Tietze, L. F.; Brand, S.; Pfeiffer, T.; Antel, J.; Harms, K.; Sheldrick, G. M. *J. Am. Chem. Soc.* **1987**, *109*, 921.

146. See footnote 3 in reference 145b.

147. (a) Mukaiyama, T.; Iwasawa, N. *Chem. Lett.* **1981**, 29. (b) Mukaiyama, T.; Tsuji, T.; Iwasawa, N. *Chem. Lett.* **1979**, 697.

148. (a) Pyne, S. G.; Hensel, M. J.; Byrn, S. R.; McKenzie, A. T.; Fuchs, P. L. *J. Am. Chem. Soc.* **1980**, *102*, 5962. (b) Pyne, S. G.; Hensel, M. J.; Fuchs, P. L. *J. Am. Chem. Soc.* **1982**, *104*, 5719.

149. Pyne, S. G.; Spellmeyer, D. C.; Chen, S.; Fuchs, P. L. *J. Am. Chem. Soc.* **1982**, *104*, 5728.

150. Yoshioka, M.; Nakai, H.; Ohno, M. *J. Am. Chem. Soc.* **1984**, *106*, 1133.

151. Hirama, M.; Uei, M. *J. Am. Chem. Soc.* **1982**, *104*, 4251.

152. Keck, G. E.; Kachensky, D. F. *J. Org. Chem.* **1986**, *51*, 2487.

153. Davidson, A. H.; Floyd, C. D.; Jones, A. J.; Myers, P. L. *J. C. S. Chem. Commun.* **1985**, 1662.

154. (a) Nicolaou, K. C.; Papahatjis, D. P.; Claremon, D. A.; Dolle, R. E. *J. Am. Chem. Soc.* **1981**, *103*, 6967. (b) Nicolaou, K. C.; Papahatjis, D. P.; Claremon, D. A.; Magolda, R. L.; Dolle, R. E. *J. Org. Chem.* **1985**, *50*, 1440.

155. Roush, W. R.; Peseckis, S. M.; Walts, A. E. *J. Org. Chem.* **1984**, *49*, 3429.

156. Boeckman, R. K.; Enholm, E. J.; Demko, D. M.; Charette, A. B. *J. Org. Chem.* **1986**, *51*, 4743.

157. Kitahara, T.; Kurata, H.; Matsuoka, T.; Mori, K. *Tetrahedron* **1985**, *41*, 5475.

158. Caine, D.; Stanhope, B. *Tetrahedron* **1987**, *43*, 5545.

159. Denmark, S. E.; Sternberg, J. A. *J. Am. Chem. Soc.* **1986**, *108*, 8277.

160. Schreiber, S. L.; Meyers, H. V.; Wiberg, K. B. *J. Am. Chem. Soc.* **1986**, *108*, 8274.

161. (a) Thomas, E. J.; Whitehead, J. W. F. *J. C. S. Chem. Commun.* **1986**, 724. (b) Thomas, E. J.; Whitehead, J. W. F. *J. C. S. Chem. Commun.* **1986**, 727.

162. Corey, E. J.; Magriotis, P. A. *J. Am. Chem. Soc.* **1987**, *109*, 287.

163. Roush, W. R.; Coe, J. W. *Tetrahedron Lett.* **1987**, *28*, 931.

164. (a) Marshall, J. A.; Grote, J.; Audia, J. E. *J. Am. Chem. Soc.* **1987**, *109*, 1186. (b) Marshall, J. A.; Shearer, B. G.; Crooks, S. L. *J. Org. Chem.* **1987**, *52*, 1236.

165. Roush, W. R.; Riva, R. *J. Org. Chem.* **1988**, *53*, 710.

166. (a) Henin, J.; Massiot, G.; Vercauteren, J.; Guilhem, J. *Tetrahedron Lett.* **1987**, 28, 1271. (b) Henin, J.; Massiot, G.; Vercauteren, J. *Tetrahedron Lett.* **1987**, 28, 3573.

167. Stork, G.; Saccomano, N. *Tetrahedron Lett.* **1987**, *28*, 2087.

168. Herczegh, P.; Zsely, M.; Szilagyi, L.; Bognar, R. *Tetrahedron Lett.* **1988**, *29*, 481.

169. Gibbs, R. A.; Okamura, W. *J. Am. Chem. Soc.* **1988**, *110*, 4062.

NONCONVENTIONAL REACTION CONDITIONS:
ULTRASOUND, HIGH PRESSURE, AND MICROWAVE HEATING IN ORGANIC SYNTHESIS

Raymond J. Giguere

OUTLINE

Organic Synthesis: Theory and Application, Vol. 1, pages 103–172.
Copyright © 1989 by JAI Press Inc.
All rights of reproduction in any form reserved.
ISBN: 0-89232-865-7

I. PREFACE

This review examines three nonconventional methods used in organic synthesis: ultrasound, high pressure, and microwave heating. Each of these techniques has made significant contributions to organic synthesis and each is at a different developmental stage. Whereas application of high pressure in organic synthesis is being presently investigated by no less than 60 research groups worldwide,[1] use of microwave heating is in its formative stages.[2]

As chemists become aware of these alternatives and explore their potential, commercial equipment will become increasingly affordable and available. In short, these techniques will join the ranks of those methods long established in the field. For example, 25 years ago the use of flash vacuum pyrolysis in organic synthesis would have been considered a nonconventional method; today its usage is sufficiently widespread for it to be considered routine.[3] One intent of this chapter is to stimulate acceptance of nonconventional methods by exposing the reader to synthetic progress that organic chemists have made in these areas.

II. ULTRASOUND

A. Introduction

The use of ultrasound to effect chemical reactions began in 1927 with investigations by Richards and Loomis involving rate studies on the hydrolysis of dimethyl sulfate and the iodine "clock" reaction.[4] Nearly a decade later, Porter and Young reported rate enhancements in the Curtius rearrangement of benzamide through the application of ultrasound.[5] In 1950, Renaud discovered ultrasound to be useful for preparing organometallic compounds such as Grignard or organolithium reagents.[6] Since 1980, the potential inherent in such pioneering efforts is being increasingly realized by a flurry of research activity in this field.

Ultrasound encompasses frequencies ranging from 20 kHz to 10 MHz with corresponding wavelengths of 7.6 to 0.015 cm. Passing ultrasound through a liquid results in acoustic *cavitation*, a phenomenon that is principally responsible for the chemical effects of ultrasound.

Acoustic cavitation is the result of alternating compressive and expansive phases of acoustic waves passing through a liquid medium.[8] Perceptually, acoustic cavitation manifests itself as a myriad of small bubbles. Under proper conditions, these bubbles undergo a cycle of generation, growth and subsequent implosive collapse known as transient cavitation.[9] Bubble dynamics of cavity growth, as well as cavity collapse, are strongly subject to subtle influences in local environment; hence the difficulty of proposing general paradigms for describing acoustic cavitation.[7a] Although the mechanism of cavitation is not completely understood, the intense energy released in the collapse phase of transient cavitation is clearly capable of effecting chemical reactions. Present-day estimates of the localized temperatures and pressures resulting from bubble collapse are 5000 K and 1000 atmospheres.[4a] Suslick *et al.* have recently first succeeded in measuring the temperature in the gas phase within the collapsing cavity (5200 K) as well as that of a thin liquid layer directly surrounding the collapsing cavity (1900 K).[10]

Given the magnitude of the localized parameters of temperature and pressure generated by transient cavitation, one might expect very large rate enhancements of chemical reactions to be observed. In fact, relatively few comparative rate studies have been conducted to determine the extent to which ultrasound enhances reaction rates.[11] In one such study, the ultrasound-promoted hydrolyses of four p-nitrophenyl esters of substituted acetic acids revealed an overall 17-fold enhancement in reaction rate, but no substrate differentiation based on substitution pattern was observed.[12] These rate enhancements are generally an order of magnitude higher than results from other comparative rate studies on homogeneous reactions.[10] As noted in this review, heterogeneous reactions appear to be more significantly affected by ultrasound than homogeneous ones. This empirical observation serves as a reminder of the sensitivity of cavitation to local environments: transient cavitation occurring near a surface significantly affects the surface and, in turn, the subsequent chemistry taking place on or near it. Unfortunately, comparative rate studies on heterogeneous systems are rare.[12] In addition, solvents play a critical role in sonication and in some cases control the observed reactions in ways that remain mysterious.[14]

Recent reviews by Suslick have been especially instructive in providing chemists with an integrated conceptual model of cavitation,[7b,c] thereby helping to bridge the gap from principally mathematical treatments of this phenomenon.[15] Readers desiring a more detailed discussion of cavitation, as well as information on reactor designs, are referred to Suslick's excellent review of these subjects.[7b,c] The following sections focus on contributions that the application of ultrasound has made to organic synthesis.

B.   Reactions Promoted by Ultrasound

1.   Conjugate Addition

The study by Luche *et al.* on the use of ultrasound (US) to promote conjugate (1,4) addition of organocopper[16] and organozinc[17] reagents to α,β-unsaturated enones has resulted in the development of sonication methodology to carry out these regioselective transformations in high yield.

Organohalides initially used to prepare organozinc compounds via sonication with lithium and zinc bromide were restricted to aryl halides.

These diarylzinc reactions, however, give results that are solvent dependent. For example, attempts to prepare ketone 1 in THF result in polymer, but the reaction occurs readily in diethyl ether.

Extension of this conjugate addition procedure to include alkyl and vinyl halides is achieved through employment of a cell disrupter type generator (Sonimasse®, 30 kHz; adjustable energy output), appropriate solvent mixtures (e.g., toluene:THF; 85:15) and the use of nickel acetylacetonate catalyst.[18] Apparatus improvements are responsible for better reproducibility and control of sonication conditions.

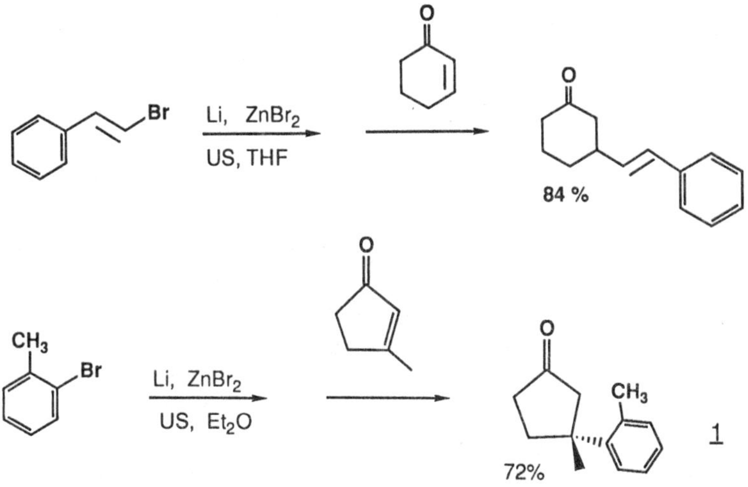

Scheme 1

This novel application is particularly attractive to synthesis because β,β-disubstituted enones are prone to undergo significant 1,2-addition with organocopper reagents under conventional methods, but work well as Michael acceptors toward ultrasonically prepared organozinc reagents. Reaction times are short and conditions mild (30–45 min; 0°C). 2-Cyclopentenone, however, failed to function well: reaction with 1-bromo-2-methylpropene resulted in 21% 3-(2-methylpropene)-cyclopentanone.[19]

In the absence of sonication, reaction times to prepare organozinc reagents increase (from 20–40 min to 1–2 h); yields decrease significantly as well (e.g., from quantitative to 25%).

A drawback is occasionally encountered when the organozinc reagent contains a hydrogen beta to the metal. In these cases 1,4-reduction of the enone moiety competes with conjugate alkyl addition, but this is usually only a complication when the enone is sterically demanding such that conjugate addition is slow relative to reduction. Side products of Wurtz-type coupling may occur, but this reaction is generally also slower than conjugate addition.[18]

This reaction has been found to occur also with enones or enals in the presence of *aqueous* solvents by sonication with zinc–copper couple.[20] Ultrasonic irradiation of zinc metal and copper(I) iodide results in immediate formation of a black suspension. Addition of alkyl iodides and Michael acceptors followed by sonication for 45 min affords 1,4-addition products in excellent yield (Scheme 3). Side reactions include only traces of 1,2-addition products; unreacted Michael acceptors are also recovered.

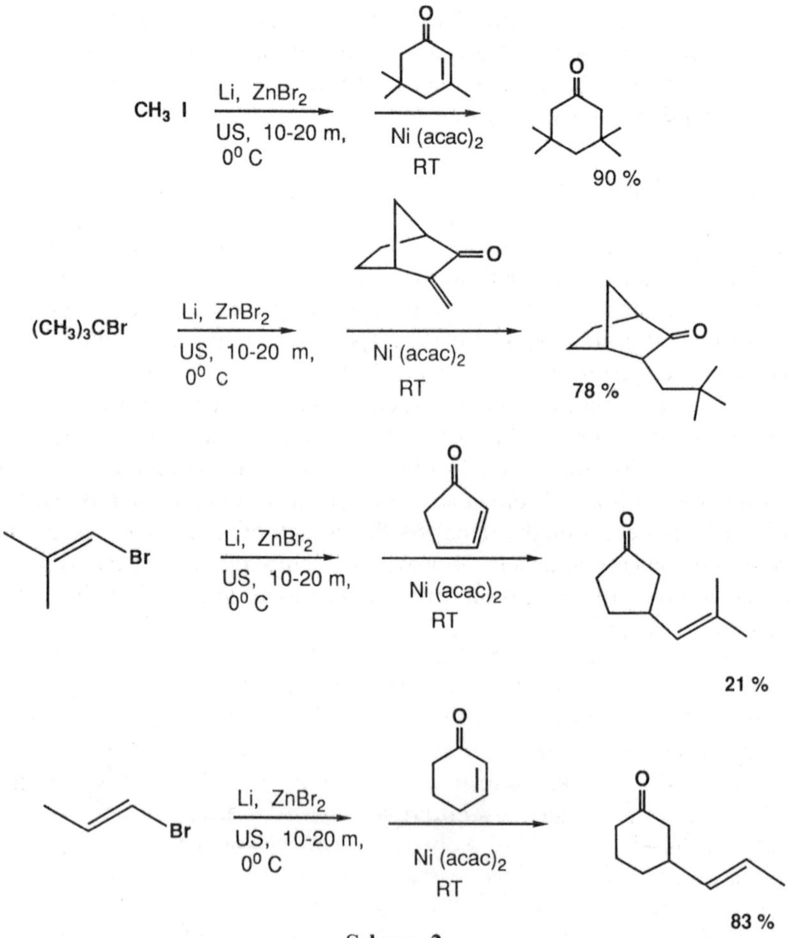

Scheme 2

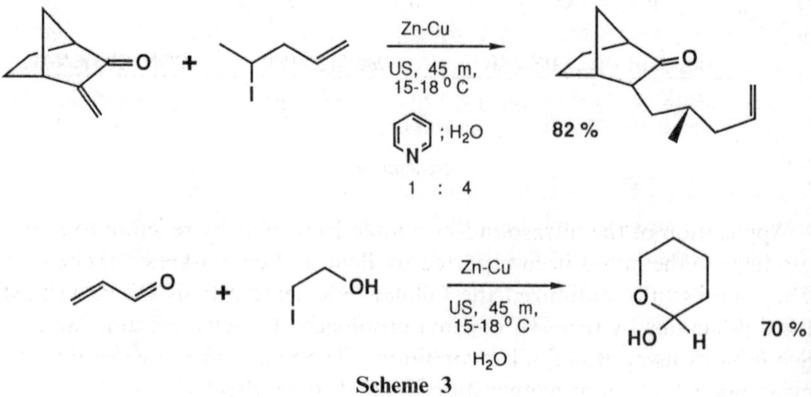

Scheme 3

The advantages of sonication methodology for conjugate addition include experimental simplicity and preparative adaptability. These sonochemically prepared reagents clearly rival the use of traditional organocopper reagents for this synthetic transformation.[21]

2. Reformatsky Reaction

Han and Boudjouk have used ultrasound to improve yields significantly as well as enhance rates of Reformatsky reactions.[22] Sonication eliminates the necessity for activated zinc, although the presence of either iodine or potassium iodide (20 or 84 mol%, respectively, based on carbonyl compound) is required for success with sonication. The reactions are carried out at 25–30°C and require 5–30 min when iodine is added, but up to 3 h with potassium iodide. These additives are believed to activate both the surface of the zinc and help suppress enolization of the carbonyl components. Solvent dependence is observed; dioxane is the solvent of choice for this reaction. Neither benzene nor diethyl ether results in high product yields, even under prolonged sonication. Scheme 4 provides a comparative example to present the advantages of sonication over conventional means for this classic reaction.

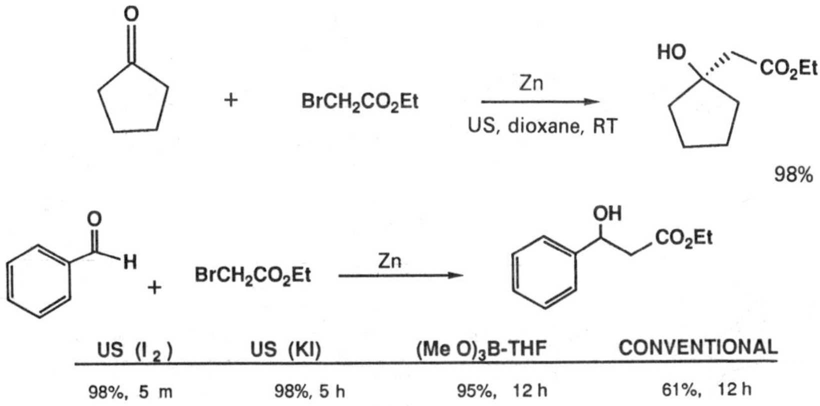

US (I_2)	US (KI)	(Me O)$_3$B-THF	CONVENTIONAL
98%, 5 m	98%, 5 h	95%, 12 h	61%, 12 h

Scheme 4

Application of the ultrasound-promoted Reformatsky reaction to natural product synthesis has been reported by Bose and co-workers[23] (Scheme 5). They successfully optimized the Gilman–Speeter synthesis of diarylsubstituted β-lactams by reaction of ethyl bromoacetate, activated zinc, and aryl Schiff bases using Boudjouk's conditions. The reactions are generally complete in 4–6 h at room temperature in good to excellent yields (70–95%).

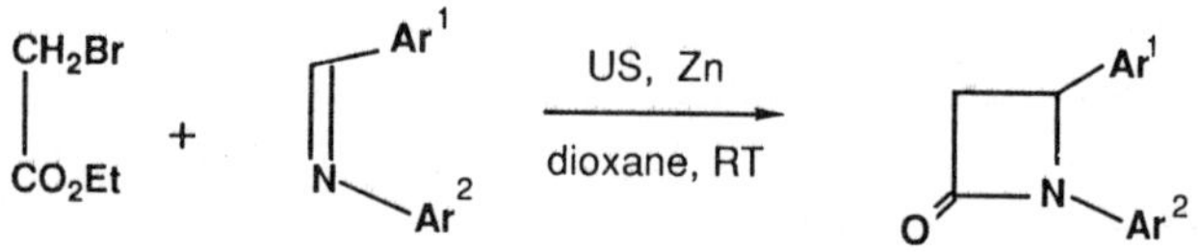

Scheme 5

Ar^1	Ar^2	Reaction time (h)	% Yield
p-$CH_3C_6H_4$	p-$CH_3OC_6H_4$	4	95
C_6H_5	p-$CH_3OC_6H_4$	5	82
C_6H_5	C_6H_5	5	70
p-ClC_6H_4	p-$CH_3OC_6H_4$	6	77

The aryl group on nitrogen can be oxidatively removed with cerium ammonium nitrate to permit derivatization of the β-lactam ring. Attempts to generalize this approach to α-bromopropionic or α-bromophenylpropionic esters were unsuccessful.

Flitsch and Russkamp have used sonication to promote intramolecular Reformatsky reactions to construct mitosane skeletons, thus initiating a pathway for syntheses of mitomycins.[24]

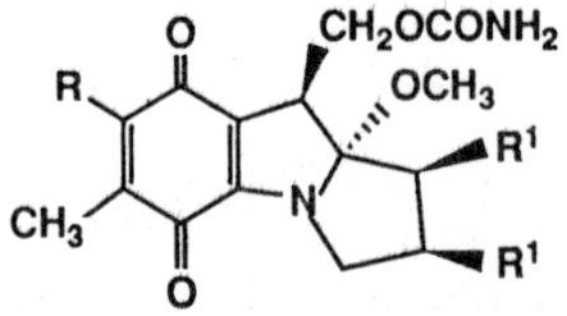

Mitomycin A: R = OCH_3, R^1 = NH_2

Mitomycin C: R = R^1 = NH_2

Scheme 6

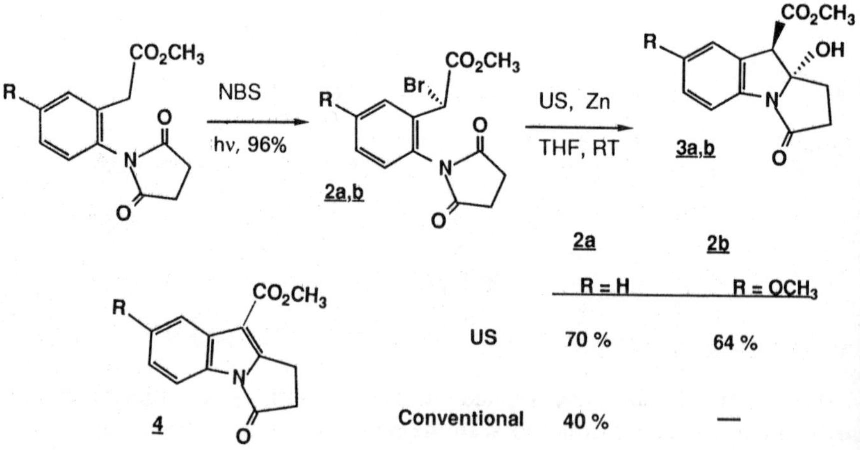

	2a	2b
	R = H	R = OCH_3
US	70 %	64 %
Conventional	40 %	—

Scheme 7

Activation of zinc in THF at room temperature for 10 min, followed by addition of **2a** or **2b** and continued sonication for 20 min results in the formation of **3a** or **3b** in 70 and 64% yield, respectively. In contrast, refluxing **2a** for 2.5 h under identical conditions affords 40% of **3a**, accompanied by an unspecified amount of dehydrated product, **4**.

Ishikawa and co-workers have applied ultrasound Reformatsky methodology to trifluoroacetaldehyde to provide pathways to trifluoromethylated allyl alcohols.[25]

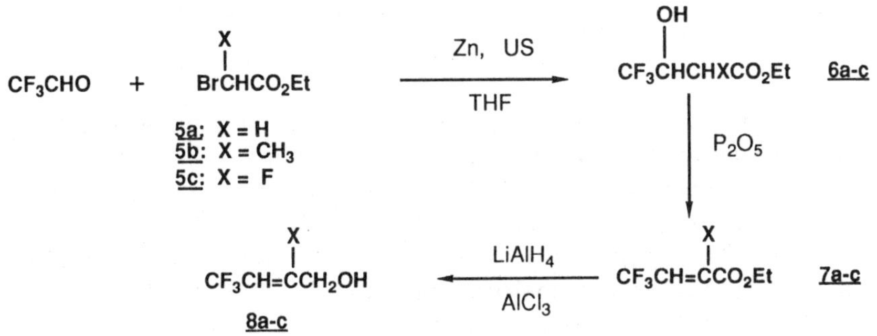

$$
\begin{array}{c}
CF_3CHO \;+\; Br\overset{\displaystyle X}{C}HCO_2Et \quad\xrightarrow[THF]{Zn,\ US}\quad CF_3\overset{\displaystyle OH}{C}HCHXCO_2Et \quad \underline{\textbf{6a-c}}
\end{array}
$$

5a: X = H
5b: X = CH$_3$
5c: X = F

$\downarrow$ P$_2$O$_5$

$$
CF_3CH=\overset{\displaystyle X}{C}CH_2OH \quad\xleftarrow[AlCl_3]{LiAlH_4}\quad CF_3CH=\overset{\displaystyle X}{C}CO_2Et \quad \underline{\textbf{7a-c}}
$$

8a-c

Scheme 8

As expected, *E* isomers are highly favored in the dehydration step (e.g., 92:08); Reformatsky yields are 52–69%. The overall transformation of **5a** to **8a** occurs in 37% isolated yield (Scheme 8). Substituting allyl bromides for the Reformatsky reagent results in the preparation of α-trifluoromethylated homoallylic alcohols in excellent yield (76–95%; Scheme 9).

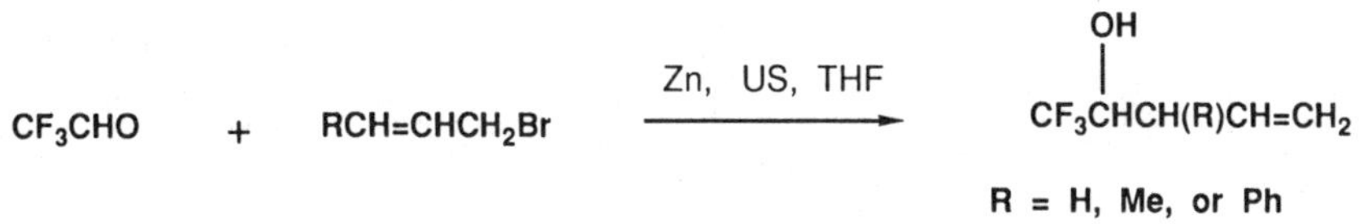

$$
CF_3CHO \;+\; RCH=CHCH_2Br \quad\xrightarrow{Zn,\ US,\ THF}\quad CF_3\overset{\displaystyle OH}{C}HCH(R)CH=CH_2
$$

R = H, Me, or Ph

Scheme 9

3. Organozinc and Organotin Reactions

Petrier and Luche have studied ultrasound-induced reactions of allyl halides and aldehydes or ketones. In the presence of zinc or tin metal, these reactions form homoallylic alcohols in *aqueous* media.[26]

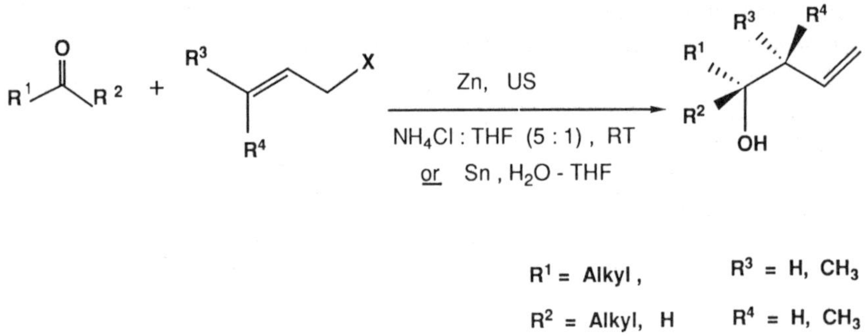

Scheme 10

Either metal may be used with equal success. The fact that these reactions can be completed in such high yield in mildly acidic media (e.g., 5:1 saturated ammonium chloride:THF) makes the possibility of *in situ* formation of organozinc or organotin reagents unlikely. Molle and Bauer have postulated that the presumed formation of organolithium reagents is not always a necessary step in the Barbier reaction.[27] More probable under these conditions is the operation of a SET mechanism.

Unfortunately, this zinc or tin variation does not enjoy the generality observed in related Barbier organolithium studies:[42] the reaction fails with alkyl or benzyl halides. On the other hand, a selectivity study using either metal revealed kinetic selectivity for aldehydes in the presence of ketones. A synthetic application is illustrated in exclusive formation of keto-alcohol **10** via regioselective addition of allyl bromide to pinonaldehyde **9**.

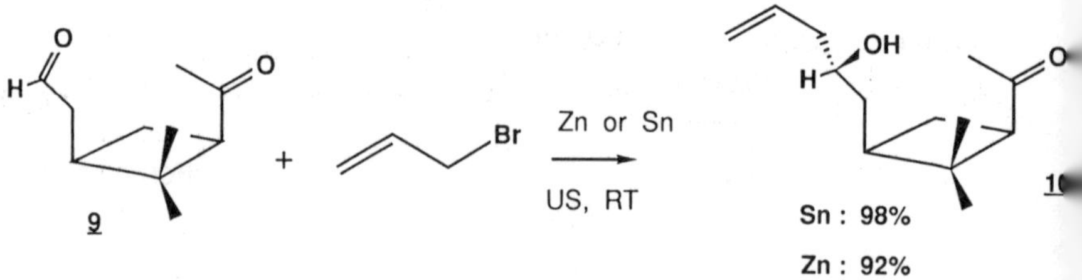

Scheme 11

Such aldehyde/ketone selectivity in aqueous media under these straightforward conditions is rare and demonstrates one of the potentials of ultrasound methodology.[28]

Knochel and Normant have prepared functionalized allyl zinc reagents and successfully added them to terminal alkynes at 45–50°C in THF to afford moderate to good yields (48–81%) of corresponding 1,4-dienes.[29] These sonicated reactions are generally complete in 2 h.

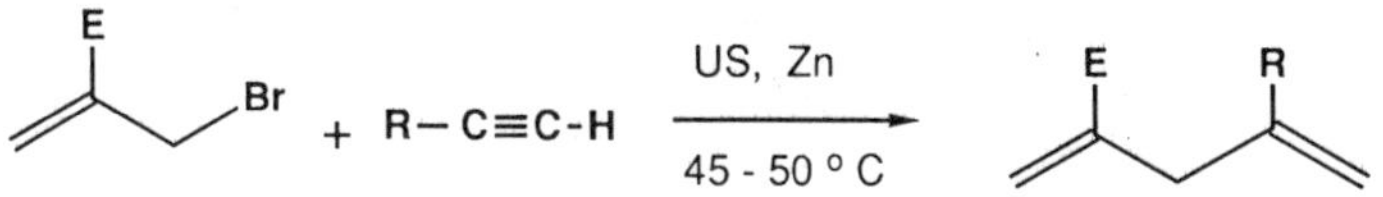

	E	R	Yield
1.	$CO_2{}^tBu$	CH_2OSiMe_3	81 %
2.	$CO_2{}^tBu$	C_4H_9	70 %
3.	CO_2Et	$CH_2CH(OEt)_2$	60 %
4.	CO_2Et	CH_2OSiMe_3	68 %
5.	$PO(OMe)_2$	CH_2OSiMe_3	72 %

Scheme 12

The products are synthetically versatile as illustrated by their easy conversion to a variety of functionalized derivatives (Scheme 13).

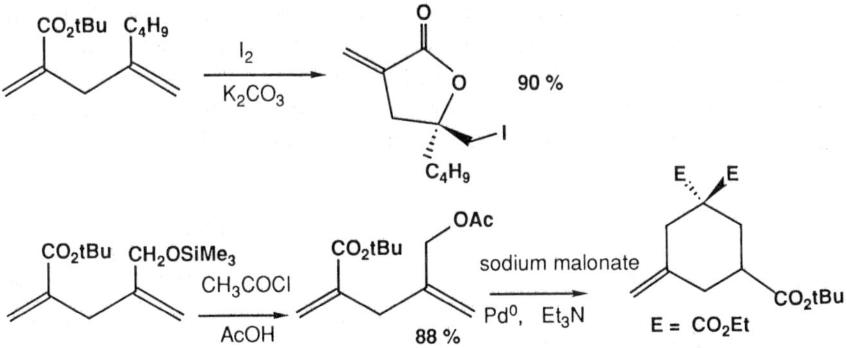

Scheme 13

Joshi and Hoffmann apply ultrasound to the Hoffmann–Noyori reaction[30] to increase yields and shorten reaction times for rapid entry to the bicyclo-[3.2.1]octene system.[31] Cyclopentadiene or furan is used as the representative diene; yields range from 75–79% and a comparative study with other reagents (e.g., diiron nonacarbonyl) used to effect these [3 + 4] allyl cation cycloadditions shows sonication conditions to be clearly superior. Reactions may be conveniently scaled up to 0.1 molar and are complete in 2 h in dioxane at 5–10°C.

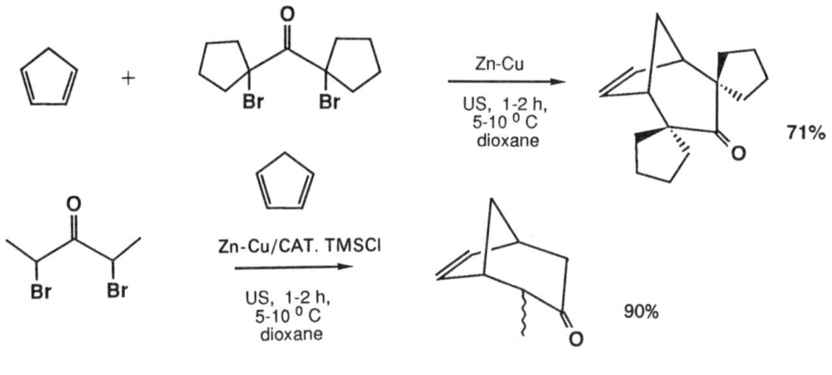

Scheme 14

4. Cyclopropanation and Methylenation

Further synthetic use of ultrasound-activated zinc is found in the Sim-mons–Smith reaction. Dichlorocarbene is also generated via sonication of *powdered* sodium hydroxide without phase-transfer catalysis. Both pro-cedures have been used to cyclopropanate alkenes in excellent yields and short reaction times. In Repic and Vogt's Simmons–Smith study, improved yields on the order of 22–55% were obtained in four cases examined.[32] A comparative kinetic study based on monitoring product formation clearly indicates that sonication is responsible for the rate enhancements observed. For example, interruption of sonication during a run caused a significant decrease in product production; reinstating sonication resulted in renewed increase in product formation. Noteworthy is the ease with which the reaction could be scaled up: experimental details are provided for cyclo-propanation of 0.5 mol α-pinene with *mossy* zinc (67%; conventional yield: 12%).

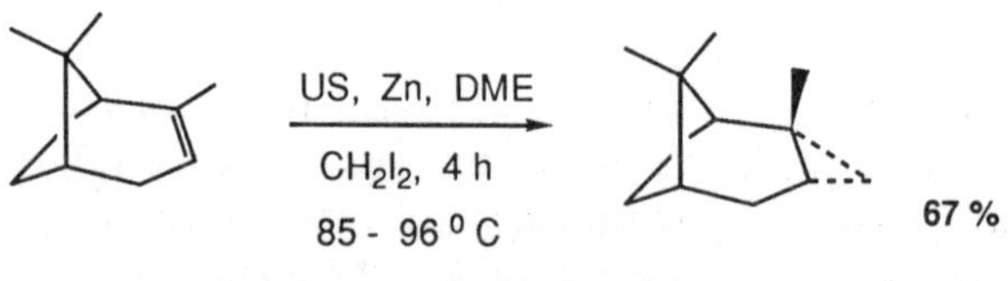

Scheme 15

A convenient sonication method for the methylenation of aldehydes with the Simmons–Smith reagent has been developed by Yamashita *et al.*[33]

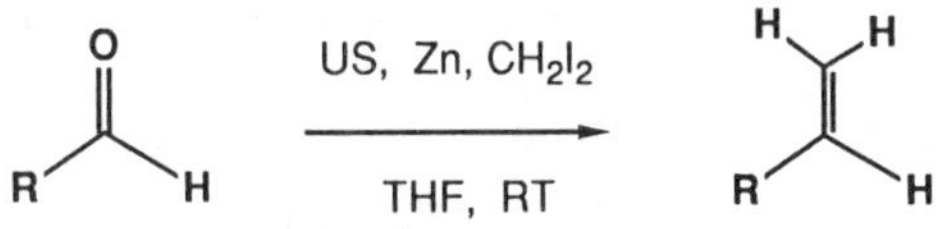

$$\underset{R}{\overset{O}{\|}}\diagdown H \quad \xrightarrow[\text{THF, RT}]{\text{US, Zn, CH}_2\text{I}_2} \quad \underset{R}{\overset{H}{\diagdown}}\diagup\underset{H}{\overset{H}{\diagdown}}$$

Scheme 16

For example, sonication of benzaldehyde, zinc powder, and diiodomethane in THF at room temperature for 20 min results in 70% conversion (by GLC) to styrene. Conversion is raised to 82% when the reaction is sonicated for 1 h. Conventional procedures require the use of activated zinc–copper couple and 6 h at 40°C.[34] Under sonication conditions, heptanal affords 1-octene in 71% yield. Attempts to apply this approach to ketones results in low yields, even after prolonged sonication (e.g., acetophenone: 5 h affords 8%).

Regen and Singh dichlorocyclopropanated 10 representative alkenes in good to excellent yield (62–99%) in 0.7–6.5 h using sonication and mechanical stirring.[35] As expected, best results are obtained with electron-rich or highly substituted alkenes; 1-hexene gives the lowest yield in the series and cyclooctene and methylstyrene the highest. These results are comparable to related reactions conducted under phase-transfer conditions and serve to teach us that sonication may be viewed as an alternative to phase-transfer catalysis.[69] Sonication is generally most successful when applied to heterogeneous or two phased systems, where surface activation and mass transfer play significant roles in reaction progress.

Success in scaling up the reactions is dependent on the power output (which is proportional to the intensity of the acoustic waves[46]) of the ultrasonicator used. Regen and Singh's work employed a 45-kHz, 35-W bath-type sonicator, which was insufficient to obtain reproducible results on reactions of greater than 5 mmol scale.[35]

5. *Magnesium and Alkali Metal Reactions*

Pioneering work by Zechmeister and Wallcave[36] and Prakash and Pandey[37] on the effects of ultrasound irradiation on cleavage of carbon–halide bonds in aromatic and aliphatic compounds laid the groundwork for initial alkali metal investigations. Considerable work using ultrasound in organic synthesis has revolved around the reaction of alkyl halides and alkali metals or magnesium *in situ*, followed by addition of carbon electrophiles to effect carbon–carbon bond formation. In this category belong the Wurtz, Grignard, Barbier, and Bouveault reactions. The success of these sonicated reactions is, in part, a result of initial rapid formation of a finely divided suspension of alkali metals in appropriate solvents.

Boudjouk and Han have dimerized representative examples of primary alkyl, aryl, benzyl, and benzoyl halides[38] as well as aryl and alkylchlorosilanes and trialkyltin halides,[39] by means of ultrasound in the presence of lithium wire in THF. These Wurtz-type reactions require relatively long irradiation times (10–60 h); yields are 50–70%. Extension of this reaction to dimesityldichlorosilane resulted in an efficient synthesis of tetramesityldisilene **11**.[40]

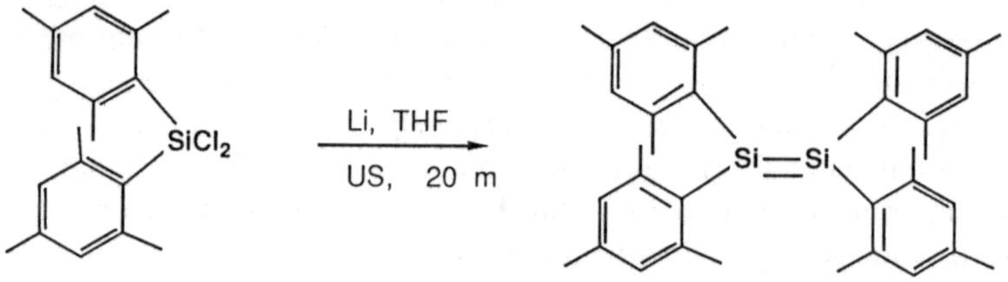

90% crude

Scheme 17

The reproducibility of this reaction, however, has been reported to be problematic.[41]

In 1980 Luche and Damiano communicated the ultrasound-induced formation of Grignard and organolithium reagents, prepared from corresponding organohalides and excess magnesium or lithium metal. Subsequent treatment of these reagents with carbonyl compounds leads to expected alcohols.[42] This key study expanded upon Renaud's early work[43] and set the stage for development of ultrasound in organic synthesis. Direct sonication of organohalides and lithium in the presence of ketones or aldehydes provides Barbier products in good to excellent yields (60–96%) in short reaction times (15–40 min).

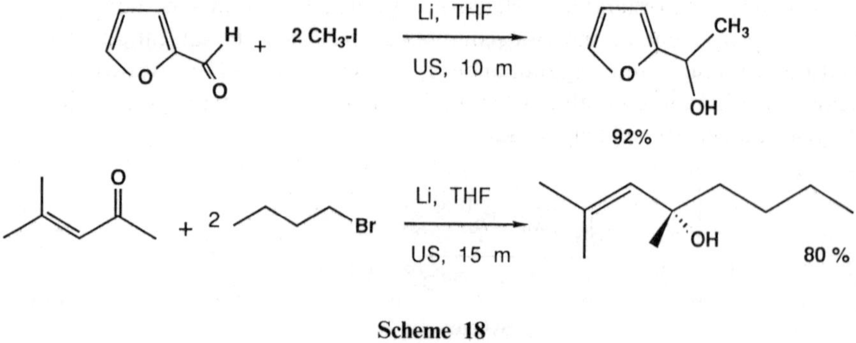

Scheme 18

These reactions can be conducted in *wet*, technical grade THF in which case 2 mol of organohalide is required. Yields increase with dry THF; this improvement requires only 1.2 molar equivalents of organohalide. Formation of organolithium reagents under aqueous conditions is not a likely step in these Barbier reactions.[27]

Ultrasound Barbier methodology is applicable to a full range of organohalides: primary, secondary, tertiary, vinyl, aryl, allyl, and benzyl all react well. This contrasts with results of conventional methods in which, for example, benzyl halides undergo predominantly Wurtz coupling under similar conditions.[44] In cases involving substituted allyl halides, carbon–carbon bond formation occurs regioselectively at the gamma position leading to products of formal S_N2' addition, as expected. Barbier products prepared via sonication of benzyl chlorides and benzaldehydes or acetophenones provide substituted stilbenes upon dehydration.[45]

More recently the ultrasound Barbier reaction of benzaldehyde, *n*-heptyl bromide, and lithium at 30 kHz has come under careful scrutiny by Luche and co-workers in order to study the role of temperature and intensity of acoustic waves on the reaction.[46] Precise kinetic analysis was not possible because sonication causes significant initial temperature jumps (as high as 15°C) in the reaction medium, presumably due to sound absorption, which complicate temperature stabilization. Direct determination of the acoustic intensity is also difficult; the authors used applied voltage as a measure of acoustic intensity. Increasing sonic intensity causes an increase in the rate of product formation, but only to an optimum value (1020 V). Increases beyond this point slow the reaction down, for reasons not entirely clear.

The initial reaction rate is fastest at 0°C, regardless of the intensity of sonication, as determined by comparative studies conducted at $-50°$, $-25°$, 0°, and 20°C. Finally, electron microscopy of the lithium surface reveals the presence of a large number of lattice defects after brief sonication (15 min), which presumably act as sites of initiation for the reaction. Simple stirring or chemical activation of lithium produces less defects and is correlated with the comparative decrease in reaction rate.

Luche and co-workers have effectively applied ultrasound methodology to the improvement of the Bouveault reaction.[47] Simple sonication of a mixture of lithium sand, organohalide, and DMF in dry THF leads to high yields of the homologous aldehyde rapidly, as exemplified in the preparation of cyclohexanecarboxaldehyde, **12**.

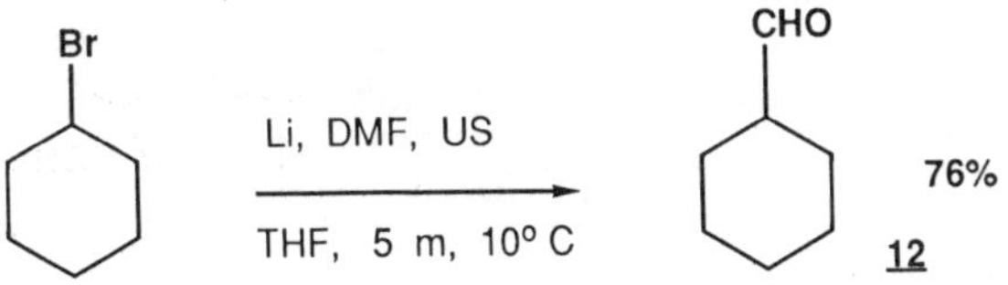

Scheme 19

Results of control studies indicate that in the absence of ultrasound, the reaction proceeds only to a small extent ($\sim 10\%$) within the same time period. The scope of the reaction is broad, extending as well to primary, tertiary, benzyl, and aryl halides. Other amides can also be used. A recent modification involving bromobenzene and N,N,N'-trimethyl-N'-formyl-ethylenediamine **13** allows subsequent regioselective metallation of the aromatic ring, thereby permitting further synthetic elaboration (Scheme 20).[48]

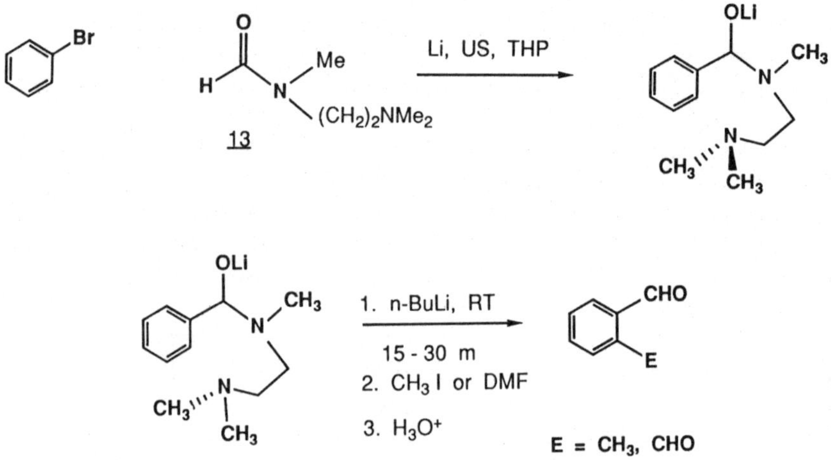

Scheme 20

The overall yield from bromobenzene and methyl iodide (or DMF) is 62%, which averages to 85% per transformation.

The ultrasound-driven Bouveault reaction, however, is reported to be frequency dependent when run in diethyl ether: **13** will not react with bromobenzene in a common 50-kHz ultrasound cleaning bath, but does when a 500-kHz wave is employed.[48a] Changing the solvent from diethyl ether to THF (or THP) permits the identical reaction to occur with 50-kHz sonication.

In 1984 Luche et al. generated colloidal potassium using ultrasound irradiation.[49] Sonication of a piece of this metal in dry toluene or xylene under an inert gas at 10°C causes rapid development of a silvery blue color, indicative of the colloid, and within minutes the potassium nugget disappears. THF may not be substituted as a solvent, nor does lithium behave similarly in any of the three solvents studied. Furthermore, colloidal sodium forms in xylene, but not in toluene or THF.

Synthetically, only colloidal potassium has been found to promote the Dieckman condensation. For example, addition of a toluene solution of diethyl adipate to colloidal potassium at room temperature results in formation of ethyl 2-oxocyclopentane carboxylate, **14** (83%), within minutes.

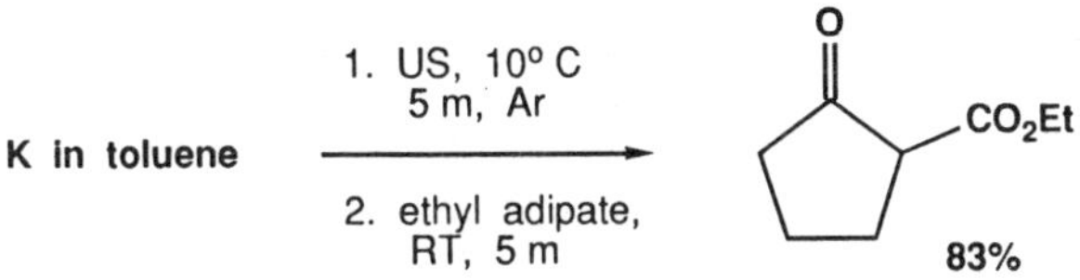

Scheme 21

The reaction is readily extended to adiponitrile (82%) and to the next lower homolog of the diester or dinitrile family with equal success. Attempted cyclization of higher homologs in order to prepare seven- or eight-membered carbocycles results only in recovery of starting materials.

Colloidal potassium is also useful for the Horner–Wittig reaction. Treatment of triethyl phosphonoacetate with colloidal potassium followed by addition of cyclohexanone affords ethyl cyclohexylidene, **15**, in excellent yield (81%).

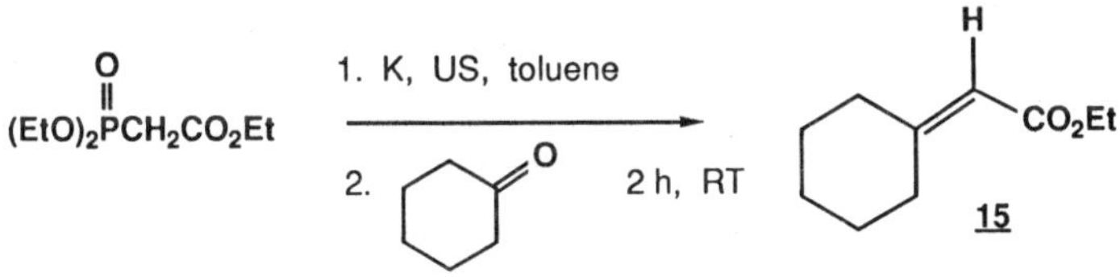

Scheme 22

Application of sonication to the barium hydroxide-catalyzed Horner–Wittig reaction of various substituted aromatic aldehydes and triethylphosphonoacetate in the presence of small amounts of water gave fair to excellent yields (32–85% by HPLC) of the corresponding *E*-alkenes.[50] Reaction times are short (10 min) and THF proved to be the solvent of choice.

6. Triorganylborane Formation

Brown and Racherla have studied the effects of sonication on hydroboration.[51] They conclude that ultrasound (US) dramatically increases the rates of many slow heterogeneous hydroborations, but only mildly accelerates the homogeneous ones. For example, the preparation of tricyclohexylborane from cyclohexene requires 24 h at 25°C under conventional conditions, but is complete in 1 h with sonication at the same temperature. Table 1 presents comparative results from this significant study.

Table 1.

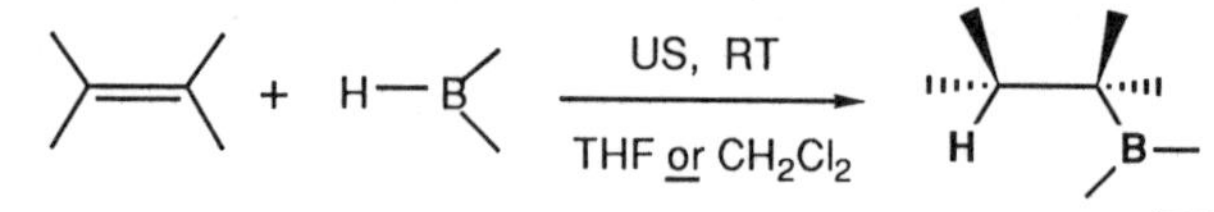

Alkene	HBR_2	Product	% Yield	US	Conventional
1. ⬡ 	HB·SMe₂	(⬡)₃B	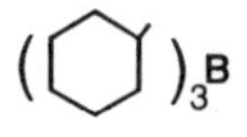 98	1 h, THF	24 h, THF
2. ⬡	 HBBr₂·SMe₂	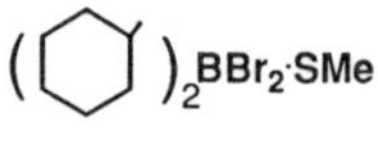 (⬡)₂BBr₂·SMe	99	1 h, CH₂Cl₂	6 h, 40°C[a]
3. ⬡ HB(⬡)₂		⬡ B(⬡)₂	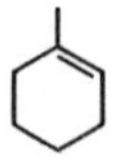 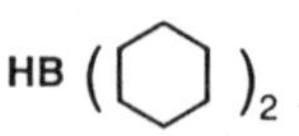 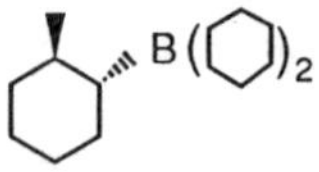 99	1 h, THF	48 h, THF
4.	 9-BBN	9-BBN	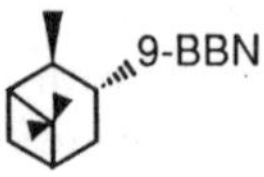 99	3 h, THF 1 h, neat	12 h, 65°C, TH
5. CH₃(CH₂)₃C≡CH	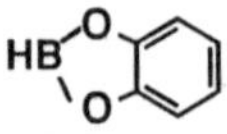 HB	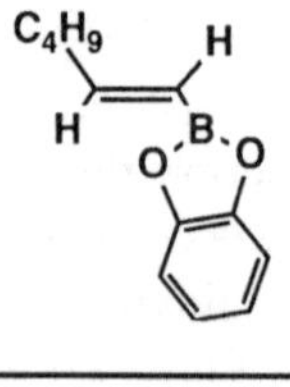 C₄H₉ H	96	6 h, neat	24 h, neat

[a] 10 mol% BBr₃ used.

More recently, Brown and Racherla have successfully employed ultrasound to prepare triorganylboranes via direct reaction of magnesium, organo-halide, and boron trifluoride etherate in diethyl ether.[52]

$$3 \text{ R-X} \xrightarrow[\text{US, Et}_2\text{O, 10 - 30 m}]{\text{Mg, BF}_3\cdot\text{OEt}_2} \text{R}_3\text{B} \quad 90 - 100\,\%$$

Scheme 23

The reaction is accelerated by ultrasound and a systematic study was conducted to compare model reactions with and without sonication. Table 2 shows representative results.

Table 2.

Organohalide	Product	GLC % Yield/Reaction Time (min)	
1. ![cyclohexyl bromide] Br	$(\bigcirc)_3 B$	99,	30
2. ![phenyl bromide] Br	$(\bigcirc)_3 B$	87,[a]	15
3. ![benzyl chloride] CH₂Cl	$(\bigcirc CH_2)_3 B$	99,	30
4. ![allyl chloride] Cl	$(\diagup\diagdown)_3 B$	94,	15

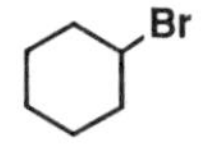

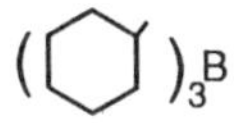

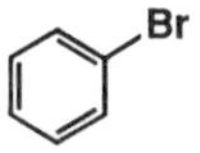

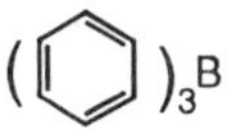

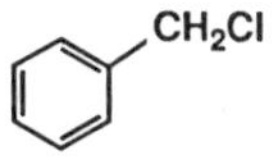

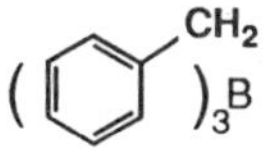

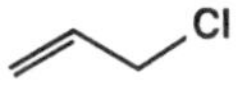

[a] Isolated yield.

n all cases examined, the ultrasound promoted reactions gave equally good
yields but significantly shorter reaction times. Investigation of the effect of
solvent, halogen, and metal revealed interesting facets of this reaction.
Different products are obtained in diethyl ether than in THF, both in the
presence or absence of sonication. Preparation of tri-n-butylborane from n-
butylbromide with magnesium gave a quantitative yield of product in diethyl
ether, but formed an ate complex in THF under identical conditions, as
revealed by ^{11}B NMR.

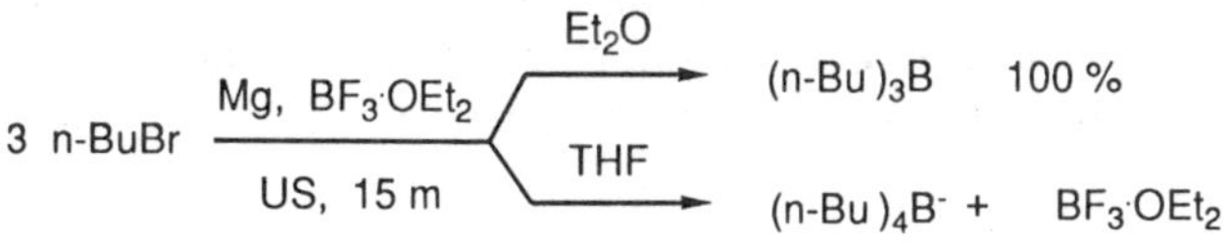

$$3\ n\text{-BuBr} \xrightarrow[\text{US, 15 m}]{\text{Mg, BF}_3\cdot\text{OEt}_2} \begin{cases} \xrightarrow{\text{Et}_2\text{O}} (n\text{-Bu})_3\text{B} \quad 100\% \\ \xrightarrow{\text{THF}} (n\text{-Bu})_4\text{B}^- + \text{BF}_3\cdot\text{OEt}_2 \end{cases}$$

Scheme 24

Using four equivalents of n-butylbromide in the THF reaction led to
quantitative formation of the ate complex. Parallel solvent behavior is
observed in pentane applying either ultrasound or conventional conditions:
formation of tri-n-butylborane is sluggish.

The order of reactivity of the butylhalides studied in this reaction is
n-BuI > n-BuBr > n-BuCl, as expected, but the iodides are prone to side
reactions, as evidenced by lower product yields. Hence, the halides of choice
for sonicated as well as conventional reactions are alkyl bromides.

Comparison of lithium and magnesium metals for the formation of tri-*n*-butylborane reveals that lithium cannot be substituted for magnesium. Sonication with lithium gave evidence of Wurtz-type products only, consistent with Boudjouk's observations.[38]

The significance of this sonicated approach is the ease with which triorganoboranes can now be prepared. Several of these compounds are not accessible by hydroboration (e.g., trimethylborane, triphenylborane, triallylborane), but are efficiently prepared by this nonconventional method. Lin and co-workers independently applied ultrasound to prepare triethylborane on preparative scale (90 mmol) via an alternative organometallic route.[53]

$$3\ EtBr\ +\ 2\ Al\ \xrightarrow{US,\ RT}\ \left[Et_3Al_2Br_3\right]\ \xrightarrow[RT]{B(OEt)_3}\ Et_3B$$

Scheme 25

These researchers found ultrasound necessary to conduct the reaction at room temperature. At 36°C the ultrasound reaction required half as much time to complete as the mechanically stirred one. They further examined the effects of initiator (iodine) concentration and ethyl bromide/alkyl borate ratios on the formation of tri-*n*-butylborane. The yield is independent of iodine concentration and highest when a 20–30 mol% excess of ethyl bromide to alkyl borate is used.

7. Perfluoroalkylation

Kitazume and Ishikawa have conducted a thorough study of the application of ultrasound Barbier methodology to the perfluoroalkylation of ketones and aldehydes.[54] The reactions are run in DMF. Product yield from ketones could be significantly improved (e.g., 18 → 41%) by the use of catalytic amounts of bis[π-cyclopentadienyl]titanium(II), which acts as a Lewis acid to activate the carbonyl group.

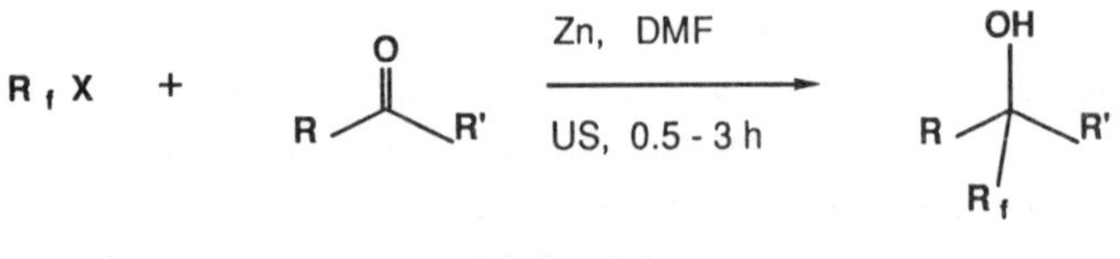

$$R_f X\ +\ \underset{R}{\overset{O}{\underset{\displaystyle \|}{R'}}}\ \xrightarrow[US,\ 0.5\text{-}3\,h]{Zn,\ DMF}\ \underset{R_f}{\overset{OH}{R\diagdown\diagup R'}}$$

Scheme 26

Further development has led these researchers to general methods for the perfluoroalkylation of vinyl and allyl positions from corresponding vinyl

nd allyl bromides by sonication with palladium(0) [Pd(PPh$_3$)$_4$] and palla-
lium(II) [Pd(OAc)$_2$] catalysts, respectively.[55] In cases involving substituted
allyl bromides, the perfluoroalkyl group was regioselectively introduced
>95%) at the gamma position, consistent with Luche's observation.[26]

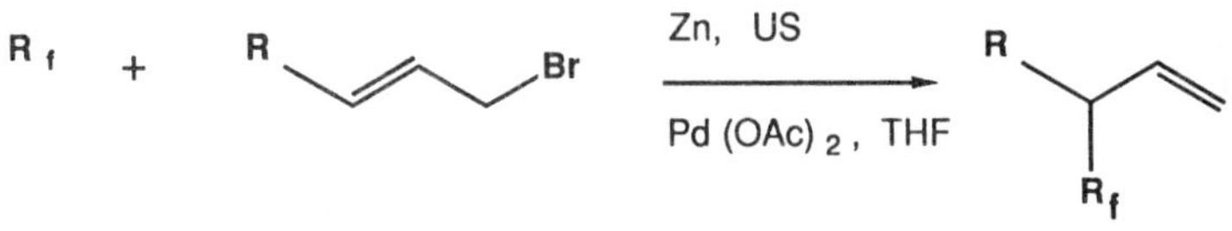

$$R_f \quad + \quad R \diagup\!\!\!\!\diagdown Br \quad \xrightarrow[\text{Pd (OAc)}_2, \text{ THF}]{\text{Zn, US}} \quad R \diagdown\!\!\!\!\diagup\diagdown$$
$$\underset{R_f}{|}$$

Scheme 27

Kitazume and Ishikawa have also hydroperfluoroalkylated terminal
lkynes by sonicating mixtures of trifluoroalkyl iodides and alkynes in the
resence of zinc and copper(I) iodide in THF.[56] Three alkyne substrates
vere examined: phenylacetylene, 1-hexyne, and propargyl alcohol; reactions
re completed in 2 h. Yields are moderate to good (52–74%) but the
eductions are not highly stereoselective, giving *E/Z* ratios of approximately
:3.

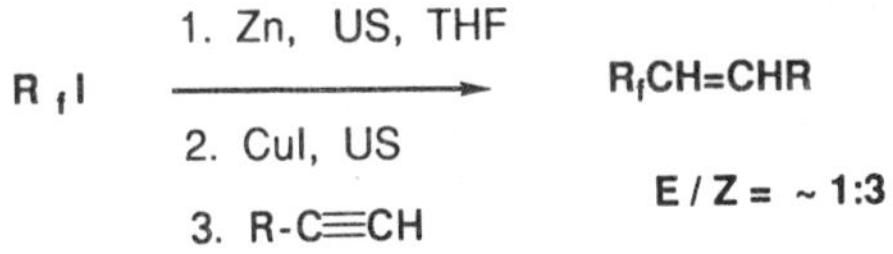

$$R_f I \quad \xrightarrow[\begin{array}{l}\text{2. CuI, US}\\ \text{3. R-C}\equiv\text{CH}\end{array}]{\text{1. Zn, US, THF}} \quad R_f CH{=}CHR$$

E / Z = ~ 1:3

Scheme 28

These researchers have furthermore demonstrated that ultrasound-
induced and Cp$_2$TiCl$_2$-catalyzed perfluoroalkylations of chiral enamines,
erived from *S*-proline and *S*-glutamic acid by the method of Enders,[57]
sult in moderate asymmetric induction in the product ketones (optical
urity: 64–76%).[54a]

Earlier work by Solladie-Cavallo and co-workers provided the first
cample of asymmetric induction with the aid of ultrasound.[58] For example,
onication of chiral chromiumtricarbonyl complex **16** and a perfluoro-
kylated iodide with zinc powder in DMF at room temperature affords,
pon acidic workup, diastereomeric mixtures of Barbier-type products. The
astereomers are easily separated by column chromatography and photo-
zed (24 h, RT) to give enantiomerically pure arylcarbinols. Asymmetric
duction is moderate: 30–66%, but the ease of diastereomeric separation
akes this approach attractive.

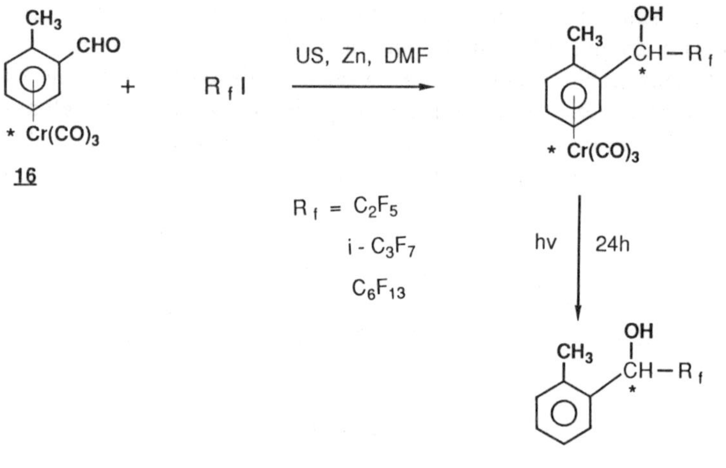

Scheme 29

8. Aromatic Anion Formation

The application of ultrasound for generation of aromatic anions was firs[t] observed by Sakurai and co-workers.[59a] A simple, practical method fo[r] rapid preparation of naphthalene lithium, anthracene sodium,[59b] an[d] biphenyl sodium consists of sonication of commercial THF mixtures o[f] corresponding aromatics and small pieces of these alkali metals. Solven[t] dependence is observed: DME can be substituted for THF, but dieth[y] ether cannot. No attempt was made to further characterize the resultin[g] anionic solutions.

Boudjouk and co-workers have studied ultrasound applications to cyclic conjugated hydrocarbons in the presence of lithium and dry THF.[60] Thei[r] results demonstrate the potential for dissolving metal reduction of anthra[-] cene, 1,3,5,7-cyclooctatetraene, and acenaphthylene to corresponding aro[-] matic dianions, evidenced by subsequent reactions of these species wit[h] electrophiles such as chlorotrimethylsilane or proton donors. For exampl[e] sonication of acenaphthylene and lithium in dry THF for 1 h at roo[m] temperature, followed by cooling to −78°C and quenching with methano[l] affords acenaphthene **17** in 90% yield.

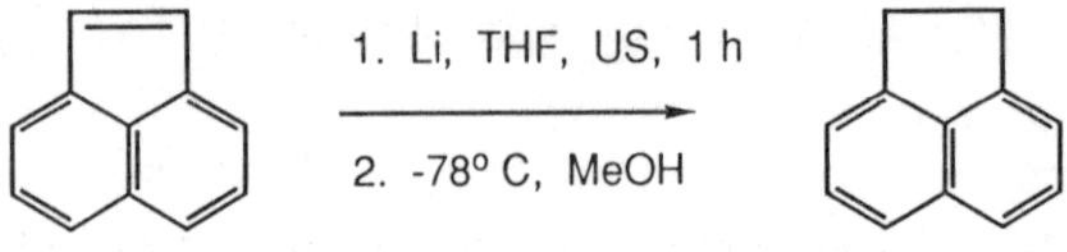

Scheme 30

Similar results are obtained when the dianion of anthracene is quenched with water to give 9,10-dihydroanthracene.

Extension of this reductive approach to α,α'-dibromo-o-xylene leads to high concentrations of monolithiated species **18**, via reductive elimination. The formation of this ionic intermediate is supported by 80% isolation of 1,5-dibenzocyclooctene on solvent removal and chromatography of the residue.

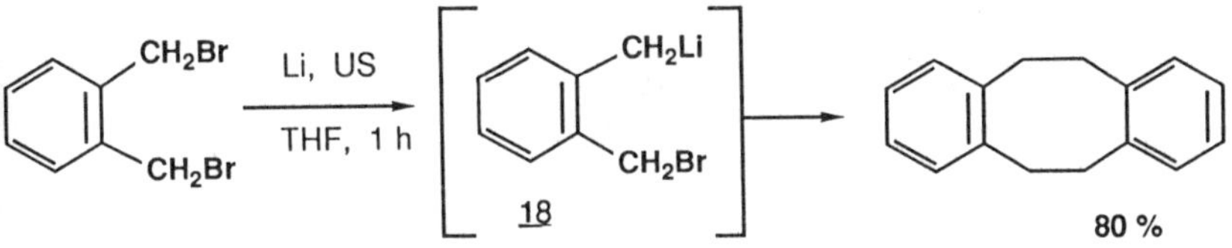

Scheme 31

This contrasts to sonication of the same compound with zinc metal, which converts readily to o-xylylene as demonstrated by the isolation of its Diels–Alder adducts on trapping with reactive dienophiles such as maleic anhydride or dimethylacetylene dicarboxylate.[61] In the absence of a dienophile polymerization occurs, and only a trace of 1,5-dibenzocyclooctene is formed. Methyl vinyl ketone and methyl acrylate can also be successfully employed, thus permitting ultrasound-induced entry into 2-substituted tetrahydronaphthalenes.

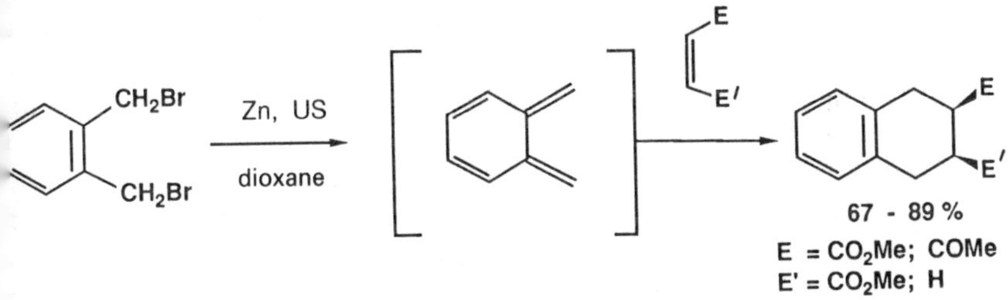

67 - 89 %
E = CO₂Me; COMe
E' = CO₂Me; H

Scheme 32

Dioxane, rather than THF, is the solvent used in the zinc-promoted reaction. Although this substitution would not be a concern under conventional conditions, sonication reactions can exhibit significant solvent dependencies. This leaves the question open as to precisely what role the solvent change may have played regarding these contrasting results.

Hydride Reduction

Han and Boudjouk have examined lithium aluminium hydride reduction of simple and deactivated aryl halides in DME.[62] Their sonication results were compared to conventional results of Brown and Krishnamurthy,[63] even though different solvents were used. In all nine cases, a significant reduction in the time necessary to complete the reactions was observed. This technique is especially attractive for reduction of electron-rich aromatic halides that are reported to be sluggish (and provide only low yields of products under conventional conditions). Table 3 presents representative data from this study.

Table 3.

$ArX \xrightarrow{LiAlH_4} ArH$		US^a % Yield/Time (h)		$Conventional^b$ % Yield/Time (h)	
1.		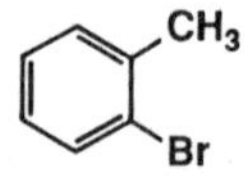 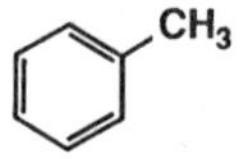 97,	5	20,	24
2.		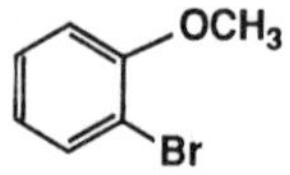 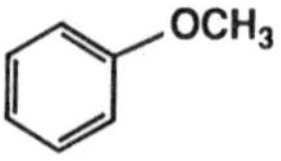 70,	7	35,	24
3.		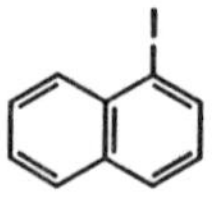 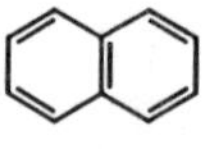 99,	6	72,	24^c
4.		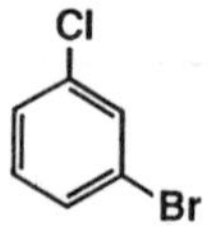 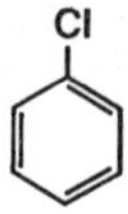 98,	4	95,	24

^{a} DME, 35°C.
^{b} THF, 25°C.
^{c} Diglyme, 100°C.

When considering these differences, bear in mind that the conventional reactions were run in solution whereas the sonication reactions were heterogeneous mixtures. Intriguing, however, is the often recurring reference to solvent dependencies with sonication reactions: "our solvent was DME, which was far more effective than THF in the presence of sonic waves."[62, 14]

10. Catalytic Reduction

Boudjouk and Han have developed an effective method of hydrogenating alkenes by sonication with formic acid/ethanol in the presence of Pd/C.[64]

The room temperature conditions are mild, reduction times short (1 h), workup simple, and yields essentially quantitative, as Table 4 discloses.

Table 4.

Alkene		Hydrocarbon; % Yield	
1.	Limonene		95
2.	α-Methylstyrene		100
3.	1,3-Cyclohexadiene	Cyclohexane	95
4.	$C_6H_5CH=CHCCH_3$ (O)	$C_6H_5CH_2CH_2CCH_3$ (O)	100
5.	$n\text{-}C_4H_9OCH=CH_2$	$n\text{-}C_4H_9OCH_2CH_3$	93

Reaction: alkene + Pd/C, US; HCO$_2$H, EtOH → hydrocarbon.

Sonication is not necessary for the successful application of this catalytic system, but does reduce the time necessary to complete the reduction. Alternatively, the reactions may be refluxed for 15 min. These conditions appear additionally suitable for hydrogenation of cyclopropane rings as well as reduction of alkynes. Cyclopropylbenzene is quantitatively converted to *n*-propylbenzene and diphenylacetylene to bibenzyl, although both heating and sonication are required in the latter case.

Further studies by Han and Boudjouk have brought a synthetically useful platinum-catalyzed reaction into focus.[65] Sonication of a variety of hydrosilanes (triethyl-, trichloro-, dichloromethyl-, triethoxy-) and model alkenes (1-hexene, 4-methyl-1-pentene, styrene, 2-methyl-1-pentene) with Pt/C catalyst results in rapid (1–2 h) hydrosilation. Yields are good to excellent (70–96%), and the catalyst may be recycled easily, especially when the products can be distilled out of the reaction mixture. In addition, no polymeric side products are produced, which, when present, complicate catalyst recovery.

Moreover, the conditions of atmospheric pressure and 30°C are the mildest heterogeneous ones known for such transformations. One example of alkyne reduction is also reported: phenylacetylene and trichlorosilane are stereo-selectively converted in near quantitative yield to *trans*-(trichlorosilyl)-styrene under these conditions. In contrast, little ($< 5\%$) hydrosilation of these substrates occurs under control conditions of vigorous agitation, even after 10–48 h.

Suslick and Casadonte recently communicated that ultrasonic treatment of nickel powder greatly enhances its usage as a catalyst for hydrogenation of alkenes. Rate increases of greater than 10^5 are reported, largely indepen-dent of alkene structure.[66] Pretreatment of nickel powder via sonication for 1 h provides optimum activity, comparable to Raney nickel, but Suslick nickel is more selective because it does not hydrogenate carbonyl groups of ketones or aldehydes.

The activity of sonicated nickel is not due to an increase in surface area of the catalyst, but rather to drastic changes in surface morphology, particle aggregation, and thickness of the surface oxide layer as revealed by electron microscopy. The latter aspect is particularly significant: the initial oxide layer is ~ 250 Å thick (Ni/O ratio $= 1.0$). After sonication for 1 h the thickness of the layer is reduced to less than 50 Å, with an Ni/O ratio of 2.0. The oxide layer reestablishes itself after only brief exposure to air (15 min); the origin of the increased catalytic activity is believed to be facile reduction of the oxide layer through sonication.

11. Nucleophilic Substitution

Ando and co-workers have conducted sonication pilot studies on both nucleophilic, aliphatic, and acyl substitution. Sonication of primary alkyl, allyl, and benzyl halides with potassium cyanide and alumina in wet toluene at 50°C results in formation of the corresponding cyanides in moderate to good yields (44–92% by GLC; best isolated: 78%).[67] Reaction times vary broadly from 6 to 85 h. As expected, alkyl bromides generally react faster than the corresponding chlorides. Best yields are obtained in the benzyl halide series; primary alkyl cyanide yields drop off as the length of the alkyl chain increases. Allyl bromide is converted to allyl cyanide in 55% yield. Surprisingly, substitution of cyclohexane, hexane, or chlorobenzene for toluene has little effect on product yield or reaction time.

Several results were contrasted to phase-transfer catalyzed controls employing 18-crown-6 or *n*-tetrabutylammonium chloride. The sonication results are consistently better than the phase-transfer ones in terms of yield achieved and lower temperatures required in these S_N2 reactions.

Aromatic acyl cyanides are prepared by sonication of corresponding acid chlorides with powdered potassium cyanide in acetonitrile at 50°C.[68] Typical yields are 70–80% in reaction times of 1–2 h. Solvent choice is critical: benzyl cyanide is prepared in 65% isolated yield in 1.5 h in acetonitrile but minimally produced (GLC: 7%) after 24 h of sonication in toluene. The toluene reaction, however, can be improved to 70% conversion (GLC) by the addition of alumina.

A practical application of sonication to nucleophilic substitution is Priebe's preparation of propargyl and allyl azides as well as azidoacetonitrile from aqueous sodium azide and corresponding organohalides.[69] Scheme 33 presents representative results.

$$R-X \quad + \quad Na\,N_3 \quad \xrightarrow[\text{X = Cl, Br}]{\text{US, H}_2\text{O}} \quad R\,N_3 \quad + \quad NaX$$

Organohalide	Azide	% Yield	Reaction Time (h)
1. HC≡CCH$_2$Br	HC≡CCH$_2$N$_3$	80	1.5
2. N≡CCH$_2$Cl	N≡CCH$_2$N$_3$	64	3
3. CH$_2$=CHCH$_2$Cl	CH$_2$=CHCH$_2$N$_3$	88	3
4. CH$_2$=CClCH$_2$Cl	CH$_2$=CClCH$_2$N$_3$	81	4

Scheme 33

Yields in these two-phase reactions are generally good to excellent (64–91%) with the exception of 2,3-dibromopropene, which gives only 48% 2-bromo-allyl azide. The reactions require four equivalents of azide ion and are complete in 1–4 h at 60°C. This approach is only marginally applicable to the synthesis of alkyl azides: 1-bromopropane yields 20% propyl azide after sonication with two equivalents of aqueous sodium azide for 3 h at 65°C. In the absence of sonication, no reaction is observed. The significance of this method is use of an inexpensive azide agent and mild conditions reminiscent of phase-transfer catalysis.

Alkylation of protected isoquinoline nitrile **19** has been achieved by a combination of ultrasound irradiation and phase-transfer catalysis.[70] With the exception of ethyl bromoacetate, however, all organohalides studied were benzyl derivatives. The two-phase reaction (50% aqueous NaOH/ toluene) is catalyzed by either triethylbenzylammonium chloride or hexa-decyltrimethylammonium bromide. The reactions are stirred as well as soni-cated; control runs without sonication were also conducted. Ultrasound irradiation improves reaction yields in all but one case and shortens the time necessary to complete the alkylations from 2 h to 20 min. The reactions failed when ultrasound was used without a phase-transfer catalyst.

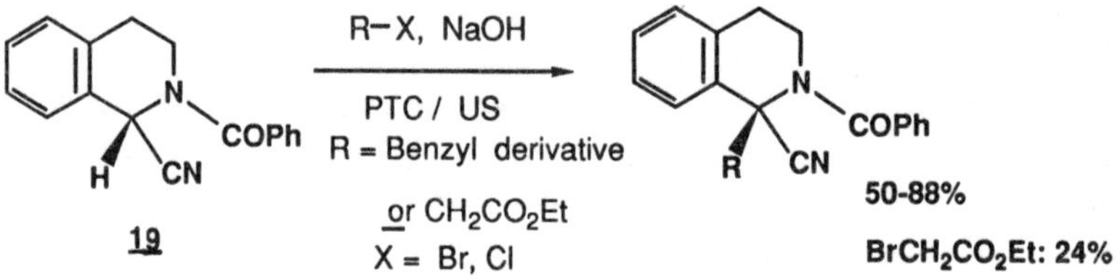

Scheme 34

A related study reports that sonication accelerates the phase-transfer reaction of amine alkylation.[71] For example, PTC methylation of indole with iodomethane and polyethylene glycol methyl ether catalyst requires 90% less time to complete when sonication is employed. In the absence of a PTC catalyst, no reaction is observed with or without sonication.

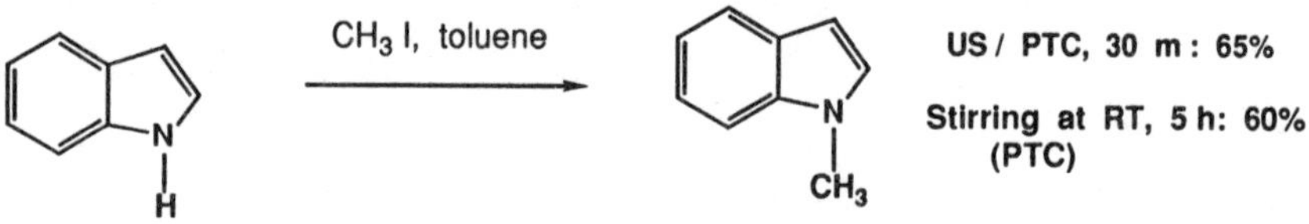

Scheme 35

Similar results are obtained with benzyl bromide and 1-bromododecane as alkylating agents. Alkylations of carbazole and diphenylamine are also significantly improved by simultaneous use of sonication and phase-transfer catalysis.

Hashimoto and co-workers use sonication to prepare spiroketones from cyclic ketones and α,ω-dibromoalkanes.[72] The method uses classic conditions (2 equiv. potassium t-butoxide, benzene), but with sonication does not require heating to 80°C. As a result, spiroalkylation of cyclopentanone (20 mmol) with 1,4-dibromobutane can be achieved at 40°C in 70% yield in 6.5 h. Replacing sonication with mechanical stirring under control conditions affords only 14% of the spiroketone. Under the 80°C conventional conditions this reaction fails because cyclopentanone self-condenses faster than it alkylates.[73] Similar results are obtained upon sonication of 1,5- and 1,6-dibromoalkanes with cyclopentanone or cyclohexanone.

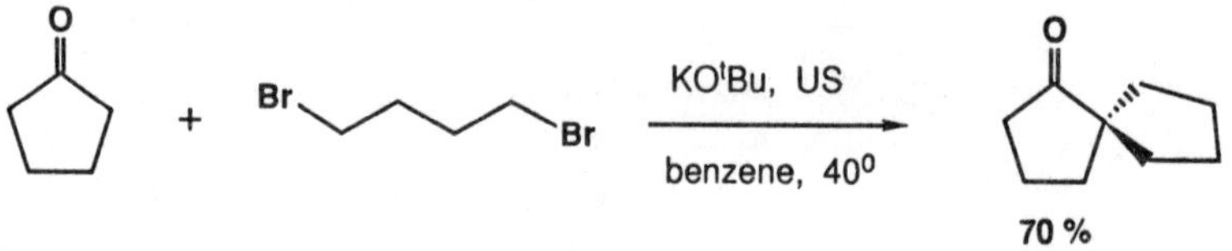

Scheme 36

12. Miscellaneous Reactions

Prolonged sonication (20–35 h) of a solution of amino-ketone **20** and potassium cyanide in glacial acetic acid with either a primary amine, ammonia, or an aniline derivative results in near quantitative conversion to corresponding aminonitriles **21a–d**.[74]

Table 5.

R	US		Conventional	
	% Yield	Time (h)	% Yield	Time (day)
21a. H	100,	25	62,	12
21b. n-Bu	88,	25	60,	12
21c. Phenyl	100,	35	73,	12
21d. Benzyl	99,	20	79,	12

This Strecker synthesis is solvent dependent: switching to water or aqueous ethanol produces cyanohydrins as sole products. By contrast, replacement of sonication by vigorous stirring results in 20–38% lower yields, even after 12 to 13 days.

Significant rate accelerations have been observed in the sonicated oxidation of secondary alcohols with powdered potassium permanganate.[75] These reactions are conducted either in hexane or benzene at 50°C; reaction times for 1 mmol runs vary from 5 to 40 h, but yields are generally very good (65–93%). Comparative studies using mechanical agitation reveals ultrasound to be responsible for the rate enhancements. Mechanical agitation exhibits a solvent dependency favoring more polar solvents, but this dependence ceases on sonication. Extending this approach to the preparation of aldehydes is not fruitful, except for preparing α,β-unsaturated enals from allyl alcohols. Noteworthy is the observation that a mixture of potassium permanganate and neutral alumina improves the rates of mechanically agitated reactions to a level comparable to that of the sonicated ones.

Einhorn and Luche have developed a practical application for the preparation of butyllithium reagents by sonication of the corresponding alkyl chlorides and lithium sand in THF at ambient temperature.[76] The reaction is complete upon disappearance of the metal (15–30 min). Initial addition of appropriate amines allows direct preparation of useful organic bases such as lithium diisopropylamide (LDA) or lithium tetramethylpiperidide. This *in situ* method is also extendable to deprotonation of Wittig reagents, 1,3-dithiane, terminal acetylenes, and sulfones, but fails to produce the dimsyl anion from DMSO.[77] Quenching the resulting anions with electrophiles (e.g., aldehydes, ketones) affords the expected products in high yields.

Additional applications of ultrasound include preparation of 3-aryl-glutaronitriles,[78] α-aminonitriles,[79] conversion of amides to thioamides,[80] reduction of α,α'-dibromoketones with metallic mercury and subsequent reaction with carboxylate anions or ketones,[81] polymerization of dichloro-silanes,[82] preparation of alkali metal selenides and diselenides,[83] acid-cata-lyzed acetylation of sugars,[84] and ligand exchange of metal carbonyls.[85] Synthesis of transition-metal carbonyl anions, which often require high temperatures and high pressures of carbon monoxide, has been achieved ultrasonically by Suslick and co-workers under much milder conditions.[86]

C. Conclusion

Since 1980 the area of sonochemistry has seen a dramatic increase in research activity and much significant synthetic progress has been made. In nearly all cases, optimization of ultrasonic conditions permits an increase in reaction yield and/or shorter reaction time for a given transformation. Herein lies much potential for further practical applications, especially with heterogeneous reactions.

In some cases, sonication produces products different from those obtained under conventional conditions, adding to the fascination and mystery of this methodology. In the preceding paragraphs, for example, we have seen repeated evidence for selection of reaction products based on solvent choice, solvent choice being crucial to the optimization of ultrasonic procedures, and solvent change leading to abrupt changes in chemical reactivity (e.g., with colloidal potassium[42]). Although little work has been done to substantiate these solvent effects, one generalization appears to hold and serves as a useful rule of thumb[14] for sonication of organic reactions: solvents possessing low vapor pressures are generally more effective than those possessing high vapor pressures.[87]

The effect of sonication on organic reactions can be extremely sensitive even to subtle changes in solvent (e.g., THF/ether,[17,59a] THF/dioxane,[60,61] or toluene/xylene[49]). This fact must be carefully considered when applying this methodology. A thorough study of the solvent effects on sonicated reactions is warranted to shed light on this important variable in sonochemical reactions.

These solvent considerations aside, the magnitude of activity and progress accomplished in this area leaves no doubt that the use of ultrasound in organic synthesis has come of age.

III. HIGH PRESSURE

A. Introduction

Application of high pressure to organic synthesis continues to develop significantly.[88] The synthetic utility of this method can be gauged, for example, by the number of total syntheses in which high-pressure methodology has played a significant role (cf. Section III, B, 5). Outstanding reviews by Matsumoto and Sera have comprehensively covered synthetic progress in this field through 1984.[89] Consequently, the following focuses on synthetic work reported since that time.

The theory of pressure effects on chemical reactions has been extensively described.[90] A detailed theoretical presentation is beyond the scope of this review, but a brief overview of some basic concepts is appropriate.

The rates of chemical reactions are conventionally increased either by increasing reaction temperature or reactant concentrations. But a reaction may also be considered in terms of activation volume, $\Delta V^{\neq}$ (expressed in cm^3/mol), defined as the difference in partial molar volume between the reactants, or initial state, and the transition state. The activation volume for a given reaction is related to the rate constant, k, by the following equation:[91]

$$\delta \ln k / \delta p = -\Delta V^{\neq} / RT$$

where p is pressure, $\Delta V^{\neq}$ is activation volume, R is the universal gas constant, and T is temperature.

Hence, pressure variation provides chemists an additional opportunity to augment reaction rates. Activation volumes for most organic reactions range from $+20$ to $-50\ cm^3/mol$.[90d] The sign and magnitude of the activation volume influence the extent that applied pressure can exert on a given reaction. For reactions in which $\Delta V^{\neq}$ is negative, the reaction rate will increase with increasing pressure; if $\Delta V^{\neq}$ is positive, the converse holds true. A negative activation volume involves a reaction in which the transition state becomes more compact as it progresses along the free energy surface; conversely, a positive activation volume requires transition state expansion.

Reaction types that would be expected to exhibit rate increases with increasing pressure have been systematically categorized.[88k,89a] Furthermore, if the rate constant for a given reaction can be determined to within 5% precision, the activation volume for the reaction can be calculated to within 1 cm^3/mol.[88k] Le Noble and others have compiled activation volume data for hundreds of reactions.[90a,c,d, 91] Knowledge of activation volume permits calculations to be made on the extent of transition state stabilization a reaction will experience.[88j]

In practice, pressures in the range of 5–40 kbar (500–4000 MPa[92]) are required to accelerate most synthetic transformations significantly. These pressures are much higher than model calculations indicate necessary.[88j] One reason for this discrepancy is that activation volume is pressure dependent and approaches zero at infinite pressure, consequently contributing less relative stabilization to the transition state with increasing pressure.[88i,k] Additional considerations that account for deviations from theoretical expectations include the consequences of physical property changes experienced by reactants and solvents under high pressure. The elevation of solvent melting points, increased solvent viscosities,[93] as well as decreases in reactant solubilities are particularly significant factors.[88i, 89] As a result, researchers have suggested that some high-pressure reactions occur in the solid state.[88j]

B.   Reactions at High Pressure

1.   Nucleophilic Substitution

Dauben et al. have applied high pressure to enhance reaction rates of triphenylphosphine and alkyl halides or sulfonates for the preparation of

Ph_3P
15 kbar,
36h, RT

92% $Br^{\ominus}$

Table 6.

R–X	Solvent[a]	Time (h)	Temperature (°C)	% Yield
1. n-$C_6H_{11}Br$	A	24	20	89
2. n-$C_6H_{11}Cl$	A	36	40	0
3. (Br)	B	36	40	55
4. (Br, CO₂Et)	A	36	20	94
5. (OMs)	B	36	20	90
6. (OMs)	B	36	20	71
7. (OTs)	A	36	20	63

[a] Solvents: A = benzene/toluene (7:3); B = acetonitrile.

phosphonium salts. Advantages of this approach are lower temperatures required and often increased yields, which are particularly significant for the preparation of phosphonium salts containing thermally sensitive functionality.[94] For example, bromoketal **22** is converted to the corresponding phosphonium salt at 15 kbar (36 h; 20°C) in 92% yield, but suffers deketalization under control conditions at 80°C.

As a result of high-pressure methodology, the conversion of an alcohol to an homologous alkene (for example, via low-temperature mesylation) can be conducted at temperatures below 20–25°C.

Jurczak and co-workers have applied high-pressure techniques to prepare chiral cryptands (cf. **23**) by a double Menshutkin-dealkylation approach.[95]

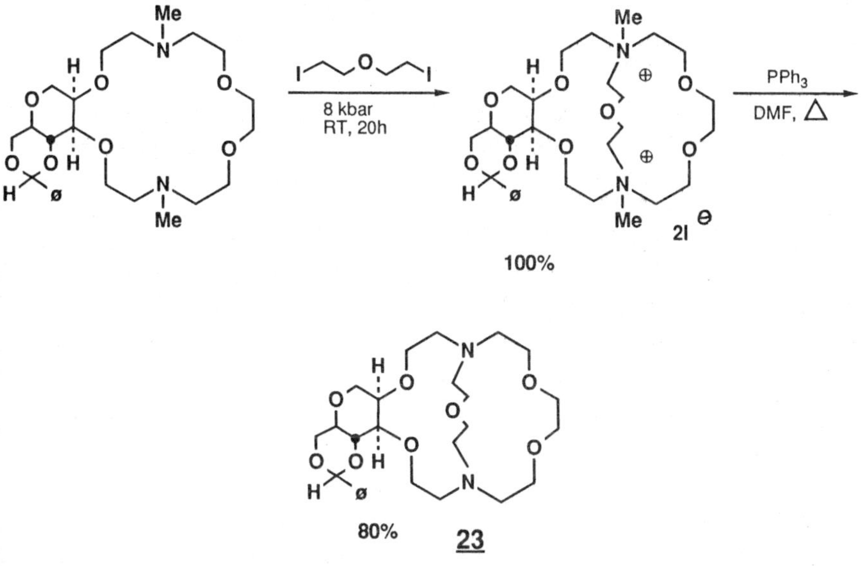

Scheme 37

The use of high pressure provides significantly better yields under milder conditions and permits the syntheses of cryptands containing sensitive functionality.

The scope and limitations of this approach with regard to the length of the bridging unit accommodatable have been studied in four diazocoronads.[96] Yields generally drop off rapidly as the length of the bridge exceeds six methylene units for these 12 to 18 atom macrocycles.

The use of high pressure to form 1,3-dioxolanes from hindered cyclic ketones and enones enhances the range of synthetic possibilities of this already widely used protecting group.[97] Dauben's thorough study involves investigation of four hindered α,α'-disubstituted ketones and three enones. Results from a study at 1 bar of the classical method for acid-catalyzed

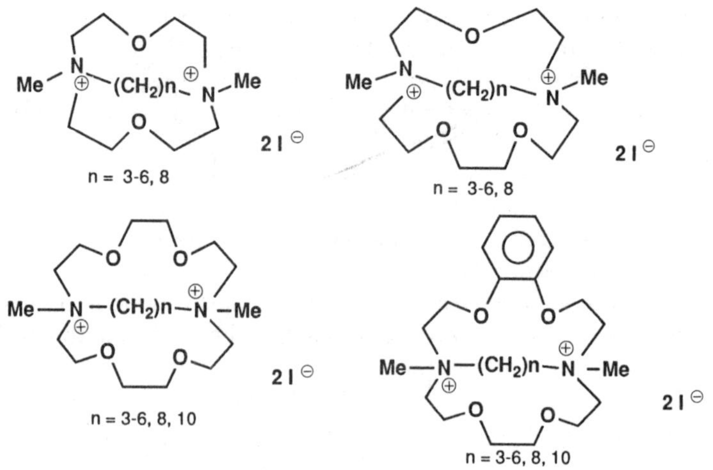

Scheme 38

ketalization (ethylene glycol, *p*-TsOH, azeotropic distillation) and the aprotic method of Noyori *et al.*[98] were compared to those from a modified acid-catalyzed procedure at 15 kbar, with triethyl orthoformate as a water scavenger. In the less hindered cases (2,2,6-trimethylcyclohexanone or camphor), high-pressure conditions did not offer significant improvements to existing methods, but did lead to better conversion under the same conditions (TsOH, 40°C, 24 h). High pressure is often successful for the formation of highly hindered 1,3-dioxolanes where classical methods fail.

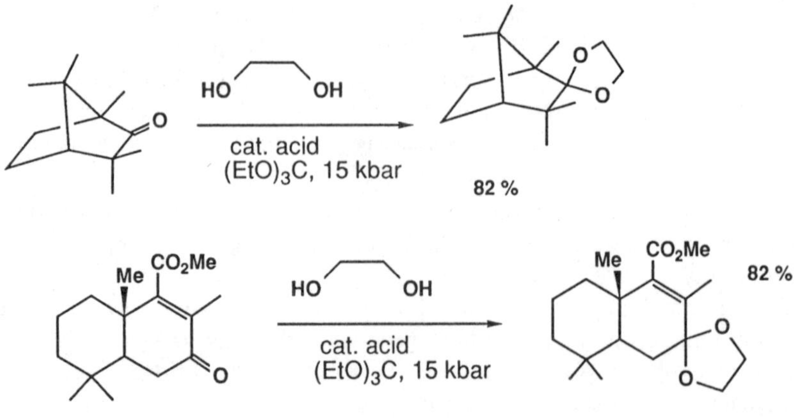

Scheme 39

High-pressure ketalization of octalone **24** epimerizes the α-methyl group, but affords 15% better yield (53%:38%).

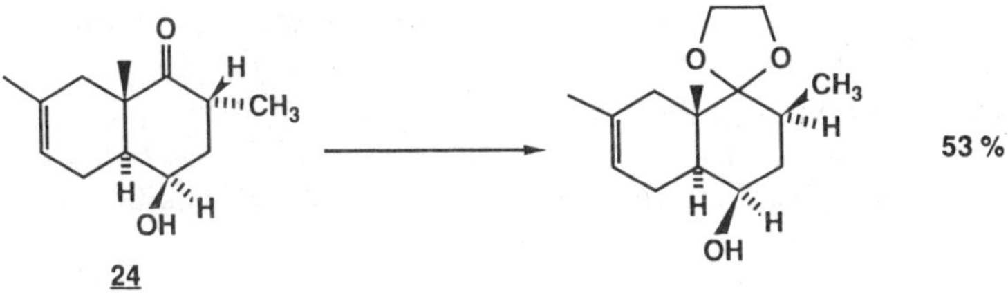

Scheme 40

53 %

High-pressure methodology is unable to effect the ketalization of 3-methyl-cyclohexenone: only complex mixtures result at 15 kbar with triethyl ortho-formate. When tri*methyl* orthoformate is substituted, methoxy derivative **25** (60%) is obtained. This product presumably results from conjugate addition of methanol to the enone prior to ketalization.[99]

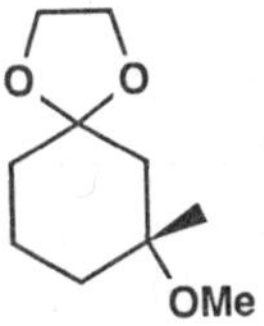

Scheme 41

Aprotic 1,3-dioxolane formation from ketones and 1,2-bis[(trimethylsilyl)-oxy]ethane and trimethylsilyl trifluoromethanesulfonate (TMSOTf) could be readily carried out in four of six compounds studied (15 kbar, 24 h, 20°C). Significantly, control runs at 1 bar ($-78°C$ for 30 min, then 20°C for 48 h) failed to produce any ketal in the same four cases. High-pressure reactions are clearly viable alternatives for the preparation of 1,3-dioxolanes from hindered ketones or enones under aprotic conditions at ambient tempera-tures. Under either set of conditions, enone **26** exhibits aberrant behavior; dimerization occurs to provide polyether **27**.

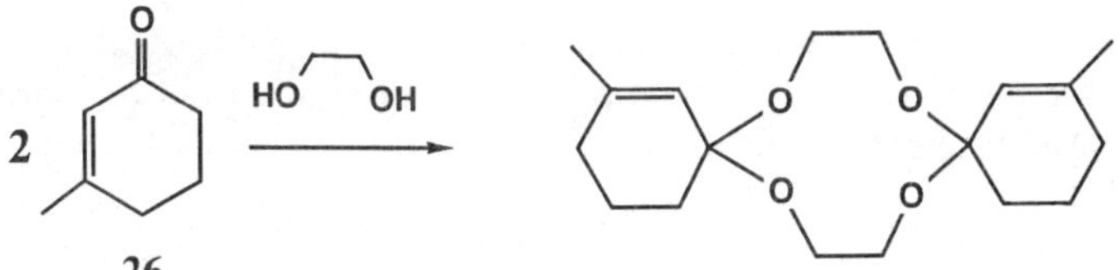

Scheme 42

In contrast to other runs, control conditions (1 atmosphere) in this case afford **27** in 27% greater yield.

The question of the magnitude of pressure necessary to produce significant rate enhancements, as well as optimize organic reactions, has been investigated for four bimolecular reactions by Dauben et al.[100] These researchers examined the following reactions from 1 to 15 kbar: alkylation of triphenylphosphine, etherification of linalool, Michael addition to bicyclo [3.3.0]oct-1,5-ene-2-one, and ketalization of fenchone.

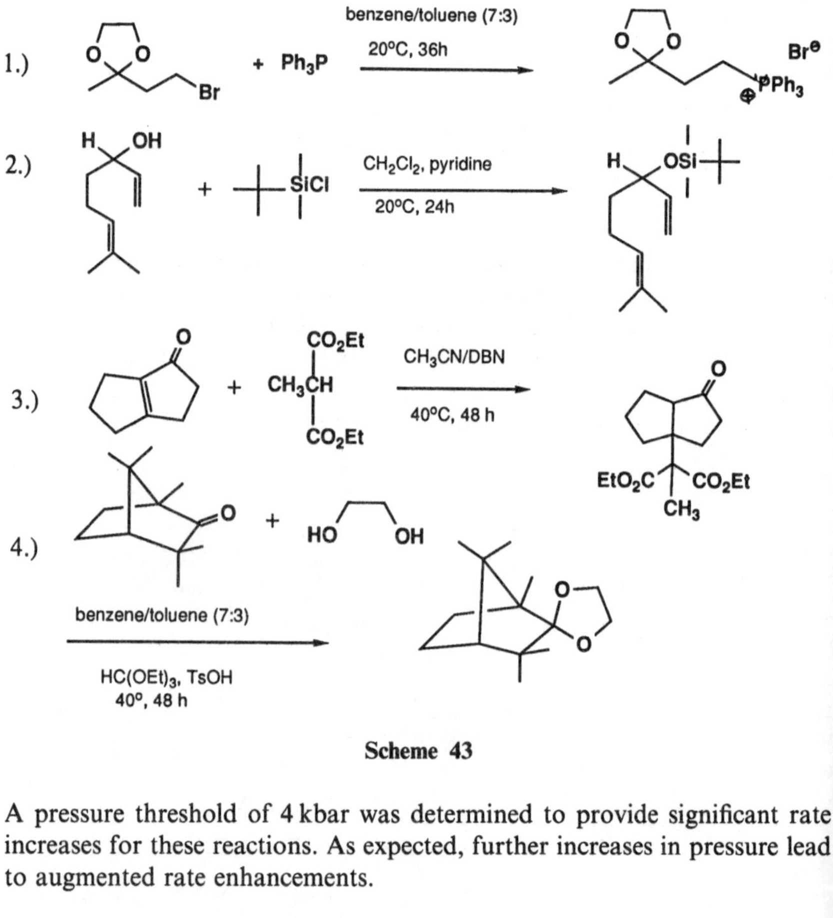

Scheme 43

A pressure threshold of 4 kbar was determined to provide significant rate increases for these reactions. As expected, further increases in pressure lead to augmented rate enhancements.

2. Nucleophilic Addition

The reduction of ketones with Alpine-Borane under high pressure at room temperature results in secondary alcohol formation with enantioselectivities approaching 100%.[101] At 6 kbar, a 15-fold increase in rate is observed compared to reactions run at atmospheric pressure. Results from this study are presented in Table 7.

Table 7.

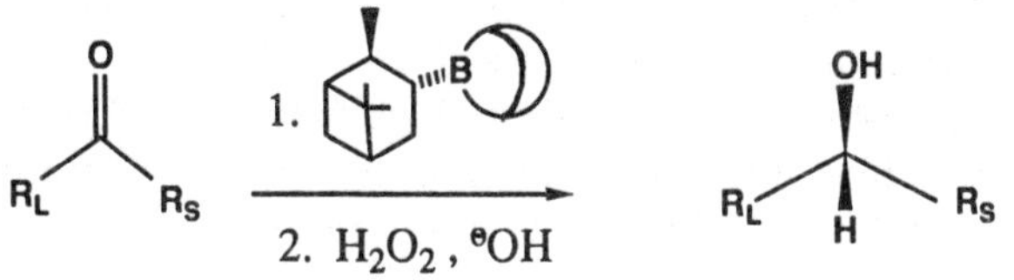
1.
2. H_2O_2, $^{\ominus}OH$
OH
H
R_L
R_S
O

Ketone	*Time*	*% ee*[a]	*Yield(%)*[b]
1. Acetophenone	1 d	92	80
2. 3-Acetylpyridine	1.5 d	92	67
3. 3-Methyl-2-butanone	1 d	82	43
4. Cyclopropylketone	5.5 d	69	65
5. 2-Octanone	<1 d	63	63

[a] When corrected for impurities in the (+)-pinene used. The values for entries 1 and 2 are 100% ee.
[b] All products possess *S* configuration.

High-pressure conditions (minimally 2 kbar) thwart competing reaction pathways of β-hydride elimination[102] and dehydroboration–reduction,[103] thus clearing the way for desired asymmetric reduction.

A boat-like transition state, based on favorable steric interactions between the pinene methyl group and the smaller of the keto-alkyl groups, is postulated to account for the absolute configurations (mostly *S*) obtained.

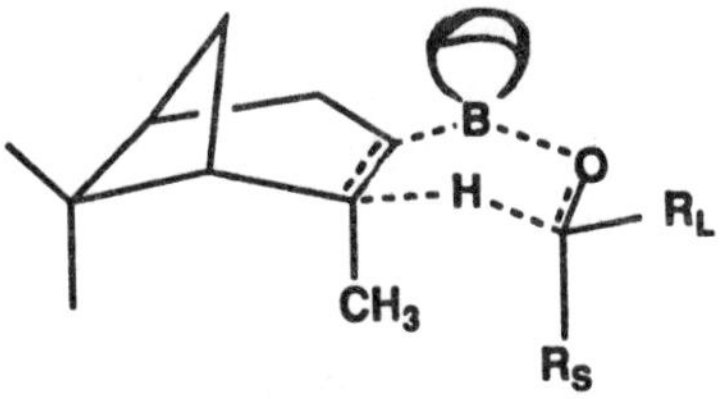
B
H
O
R_L
CH_3
R_S

Scheme 44

An enantioselective synthesis of β-amino esters has been achieved employing modified menthyl crotonates in a high-pressure Michael reaction with α-substituted benzylamines.[104]

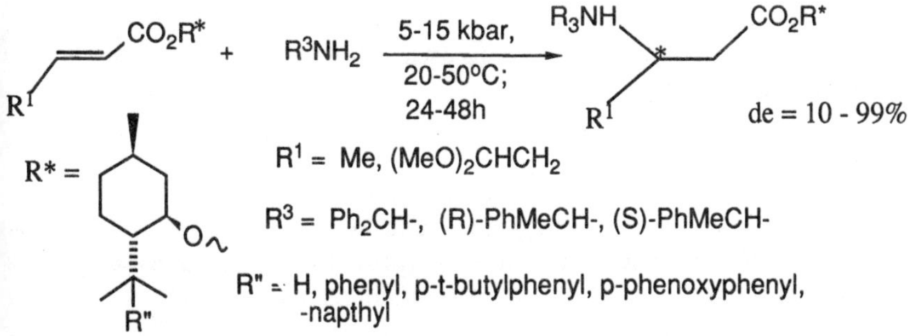
CO_2R^*
+
R^3NH_2
5-15 kbar,
20-50°C;
24-48h
R_3NH
CO_2R^*
R^1
de = 10 - 99%
R* =
R^1 = Me, $(MeO)_2CHCH_2$
R^3 = Ph_2CH-, (R)-PhMeCH-, (S)-PhMeCH-
R" = H, phenyl, p-t-butylphenyl, p-phenoxyphenyl, -napthyl
R"

Scheme 45

Diastereomeric excesses (de) dramatically increase when the methine hydrogen of the isopropyl group in the menthyl moiety is replaced with phenyl,
p-t-butylphenyl, phenoxyphenyl, or β-naphthyl substituents. 8-(β-naph-
thylmenthyl) crotonate and α-phenylbenzylamine provides a Michael product in diastereomeric excess of greater than 99%. In contrast, the same
reaction with 8-menthyl crotonate provides only 10% de, at comparable
chemical yields. These stereochemical results are rationalized by a "π-
stacking" model in which one face of the crotonate is effectively shielded
from the amine nucleophile by the aromatic ring(s). In the absence of high
pressure, a related reaction of ethyl crotonate and $(-)$-1S-phenylethylamine
(6 h, EtOH, reflux) occurs in only 28% chemical yield and with less than
10% de.[105] Apparently the additional phenyl ring plays a significant role in
effective "π-stacking." Further study will determine if these chiral esters and
high pressure are equally effective for diastereoselective Michael additions
with nonaromatic and less hindered amines.

Recently Sera and Matsumoto conducted a thorough investigation of
three high-pressure Michael reactions: (1) p-substituted thiophenols and
cyclohexenones, (2) nitromethane and chalcone, and (3) methyl 1-oxo-
indane-2-carboxylate and methyl vinyl ketone. Their study employed pressures from 9 to 15 kbar in conjunction with quinine and quinidine to effect
asymmetric conjugate addition. In general, enantiomeric excesses (ee) of 20–
50% were observed, but the application of pressure did not significantly
affect the ee ratio; in cases where it did, the percentage ee *decreased* slightly
compared to the ambient pressure results.[106] Chemical yields in all reactions
were excellent under high-pressure conditions; reaction of nitromethane and
chalcone, for example, does not occur at atmospheric pressure.

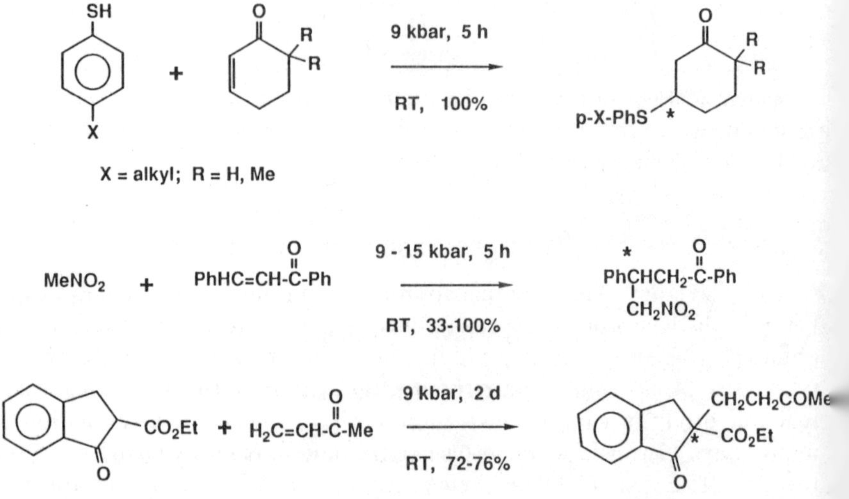

Scheme 46

3. [2 + 2] Cycloaddition

In 1983 Aben and Scheeren reported the first examples of [2 + 2] cyclo-additions conducted under high pressure.[107] At 12 kbar, monosubstituted ketene acetals and aldehydes react in dichloromethane to afford principally *trans*-oxetanes, although the cis/trans ratio is affected slightly by pressure and reaction time. Ketones undergo the high-pressure reaction only when a Lewis acid catalyst (1 mol% $ZnCl_2$) is employed. At ambient pressures this reaction occurs with simple aldehydes if a Lewis acid is used, but then *cis*-oxetanes are the major products. This reversal of selectivity is presumably a result of different transition states being involved under different conditions. The oxetane products are readily converted to β-hydroxy esters upon hydrolysis.

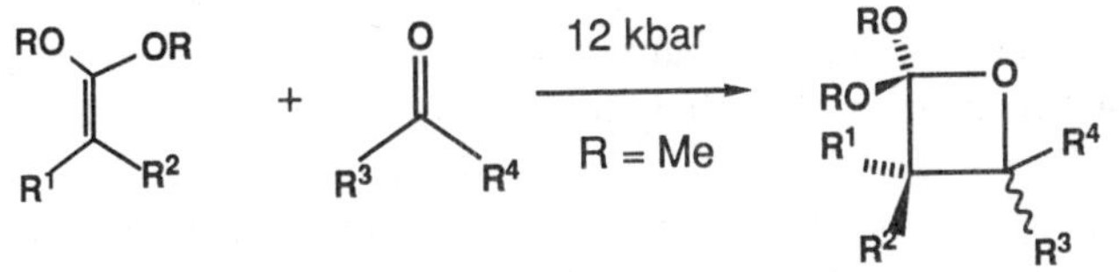

Oxetane:	R^1	R^2	R^3	R^4	Time (h)	% Yield
1.	H	H	H	C_6H_5	3	70
2.	H	CH_3	H	C_6H_5	1	90
3.	H	CH_3	H	n-C_6H_{11}	8	60
4.	H	CH_3	Cyclohexane^a		5	80

^{a} Lewis acid (1% $ZnCl_2$) required.

Scheme 47

A limited study extending this approach to α,β-unsaturated aldehydes or ketones was less promising because mixtures of up to three products were obtained.[108] Factors affecting product distribution include solvent, carbonyl type, and β-substitution pattern; a clear rationale for predicting these reaction products cannot emerge until a comprehensive investigation is undertaken. Germane to these results are reports from two groups on the reaction of related *O*-silylated ketene acetals and activated α,β-unsaturated ketones to provide exclusively products of Michael addition.[109]

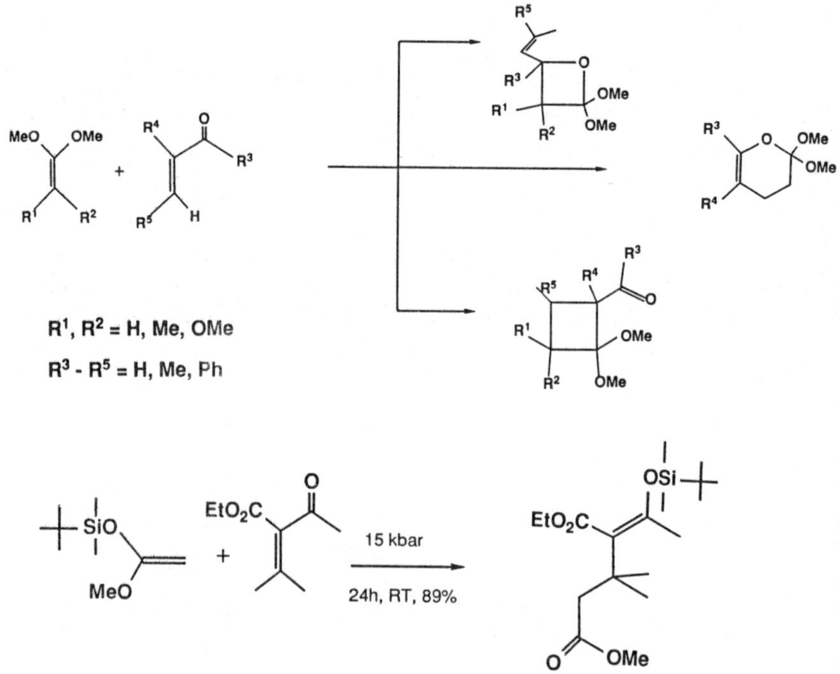

R^1, R^2 = H, Me, OMe

R^3 - R^5 = H, Me, Ph

Scheme 48

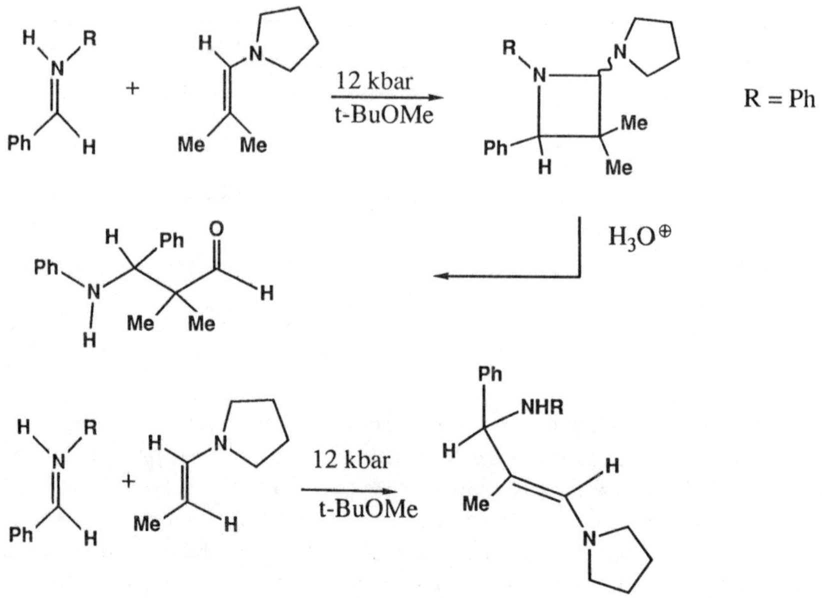

R = Ph

Scheme 49

Recently, Scheeren and co-workers have explored high-pressure reactions of imines and ketene acetals, enol ethers, or enamines.[110] This investigation resulted in the first isolation of azetidines; hitherto such compounds had been proposed as reactive intermediates. The azetidines resulting from enamines and imines are unstable at ambient pressure, but could be characterized spectroscopically. Subsequent hydrolysis of the unstable azetidines affords corresponding β-amino carbonyl compounds in fair yields (40–65%). Enamines containing β-hydrogens do not form azetidines with imines, but afford acyclic compounds directly.

Imines derived from benzaldehyde similarly add to ketene acetals, but form stable (i.e., isolable) azetidines only with tetramethoxyethene. Enol ethers containing α-hydrogens also provide stable azetidines when reacted with similar imines under high pressure at 50°C. Enol ethers bearing a methyl substituent in the α-position afforded acyclic compounds under identical conditions. Neither reaction occurs without the influence of high pressure.

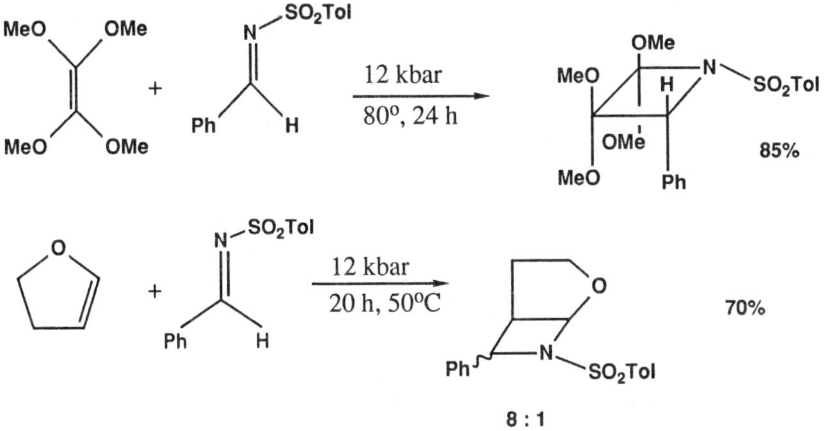

Scheme 50

The reaction of glycals with toluene-4-sulfonyl isocyanate at 10 kbar provides a stereospecific entry to β-lactams, but these [2 + 2] cycloadducts revert to starting materials upon standing at ambient temperatures and pressures.[111]

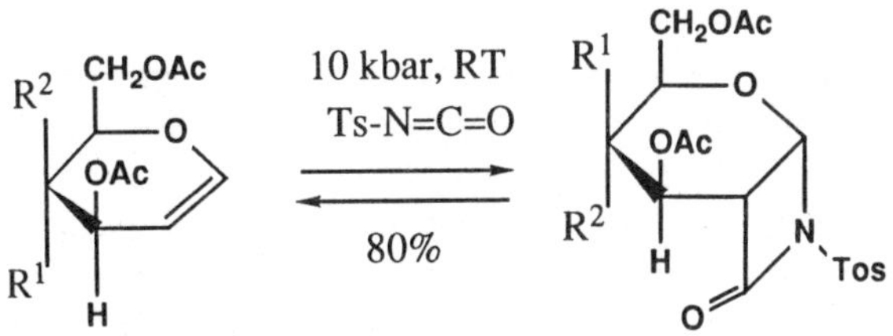

Scheme 51

4. Diels–Alder Reaction

Over a decade has passed since Dauben and Krabbenhoft's pioneering study on high-pressure Diels–Alder reactions. Facile entry to 7-oxabicyclo-[2.2.1]heptyl derivatives was accomplished by reactions of furan and enophiles bearing only one electron-withdrawing group.[112] This work laid the foundation for many subsequent investigations on applications of high pressure to Diels–Alder reactions.[113]

More recently, the cycloaddition of furan and 1,4-benzoquinone has been achieved at 22 kbar,[114] but yields are low (29–38%) and the adducts unstable at ambient pressures, quickly reverting to starting materials. The exo adduct is more stable than the endo one: cycloreversion of endo **28** is complete at $-8°C$ in 1 h; exo **29** requires 12 h at 5°C. As a result of this lability, kinetic aspects of this retro-Diels–Alder process were difficult to study. However, reaction of substituted 1,4-benzoquinones and 3,4-dimethoxyfuran provided model endo-cycloadducts on which kinetic data were obtained for analogous retro-Diels–Alder reactions.[115]

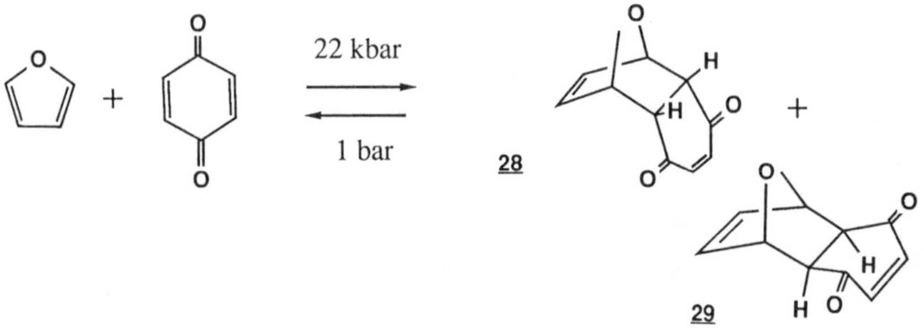

Scheme 52

These results explain why cycloadducts of furan and benzoquinones have never been prepared thermally.

The high-pressure Diels–Alder reaction of a furan-derived adduct and 2-substituted tropone derivatives **30a,b** leads to homobarrelenones, **31a,b**, after thermal cycloreversion.[116]

The low yields in the cycloreversion are due in part to the facile isomerization of **31a,b** to substituted indanones.

Pyrrole possesses greater aromatic character than furan; consequently, success with this diene in Diels–Alder reactions has been limited. Only extremely reactive dienophiles, such as tetrakis[trifluoromethyl]-Dewar thiophene,[117] undergo [4 + 2] cycloadditions with pyrrole. The more common pathway follows a Michael-type reaction to provide 2-substituted pyrroles **32**. Thermal reaction of pyrrole and dimethylacetylene dicarboxylate has been reported to provide indole derivative **33a**,[118] presumably a result of the

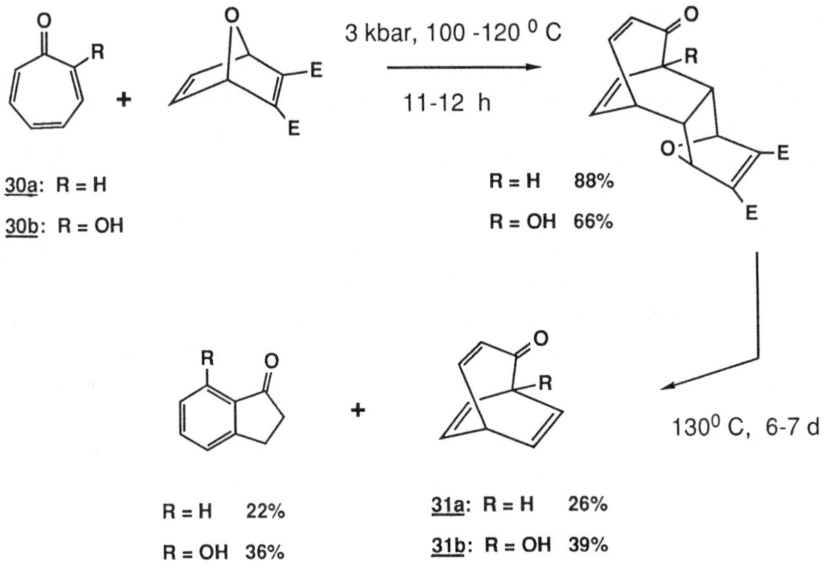
3 kbar, 100 -120 ⁰ C
11-12 h
30a: R = H
30b: R = OH
R = H 88%
R = OH 66%
130⁰ C, 6-7 d
R = H 22%
R = OH 36%
31a: R = H 26%
31b: R = OH 39%

Scheme 53

initial Diels–Alder adduct reacting with a second dienophile, but this was not confirmed by Kotsuki *et al.*[119] Conducting this reaction thermally or under high pressure affords products resulting from Michael-type addition. Surprisingly, high-pressure conditions result in lower yield.

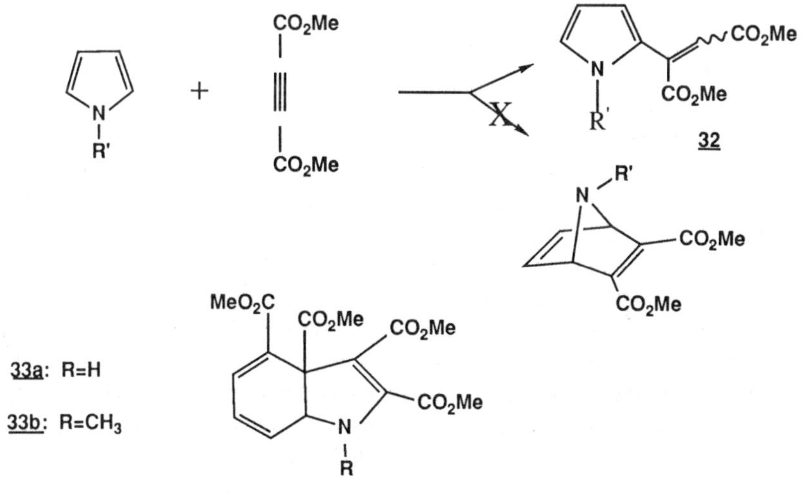
32
33a: R=H
33b: R=CH3

Scheme 54

Substituting N-methylpyrrole in the high-pressure (or thermal) reaction produces analogous indole derivative **33b** as the sole product.[120]

High pressure is often able to suppress alternative pathways in cases where electron-withdrawing groups are bonded to pyrrole at nitrogen. Isaacs and co-workers have recently conducted a study on the reaction of acylpyrroles ($R = COPh$; CO_2Et; $COCH_3$) and four dienophiles: maleic anhydride, N-methylmaleimide, N-phenylmaleimide, and dimethylacetylene dicarboxylate.[121] Selected results are presented in Table 8.

Table 8.

Pyrrole	Dienophile[a]	Solvent	Conditions	% Yield [exo/endo]
1. COPh	NPM	EtOAc	14 kbar, 90 h, 34°C	46/45
2. COPh	NPM	Benzene	12 kbar, 60 h, 25°C	80/0
3. COPh	DMAD	Neat	13 kbar, 24 h, 25°C	40
4. COPh	MA	CHCl$_3$	11 kbar, 160 h, 24°C	0/20
5. COPh	MA	EtOAc	12 kbar, 300 h, 25°C	25/0
6. CO$_2$Et	DMAD	EtOAc	12 kbar, 100 h, 24°C	35

[a] NPM, N-phenylmaleimide; MA, maleic anhydride; DMAD, dimethylacetylene dicarboxylate.

The adducts of N-phenylmaleimide are formed in significantly higher yield than those from maleic anhydride. Furthermore, the maleic anhydride adducts are labile and, in solution, undergo retro-Diels–Alder reaction in a few hours.

Solvent plays an important role in determining the endo/exo product ratio. For example, switching solvents from ethyl acetate to chloroform in the reaction of N-benzoylpyrrole and maleic anhydride results in complete reversal of endo/exo selectivity (cf. entries 4 and 5). Other examples are less dramatic; a rationale for these results has yet to be formulated. In general, endo isomers appear to be favored in less polar solvents.[122]

Quadricyclane and methyl propiolate undergo a [2 + 2 + 2] cycloaddition at 100°C under 10 kbar pressure for 24 h to afford near quantitative yield of tricyclononadienecarboxylate **34**.[123] Conducting the high-pressure reaction with a 2:1 ratio of reactants for longer periods (65 h) results in isolation of polycyclic bis-adduct **35** in 96% yield. The yield of **34** is 60% under conventional conditions (100°C, 24 h); only traces of bis-adduct **35** form. Homo-Diels–Alder reaction of norbornadiene and methyl propiolate under similar conditions gives monoadduct **36** in 75–85%.

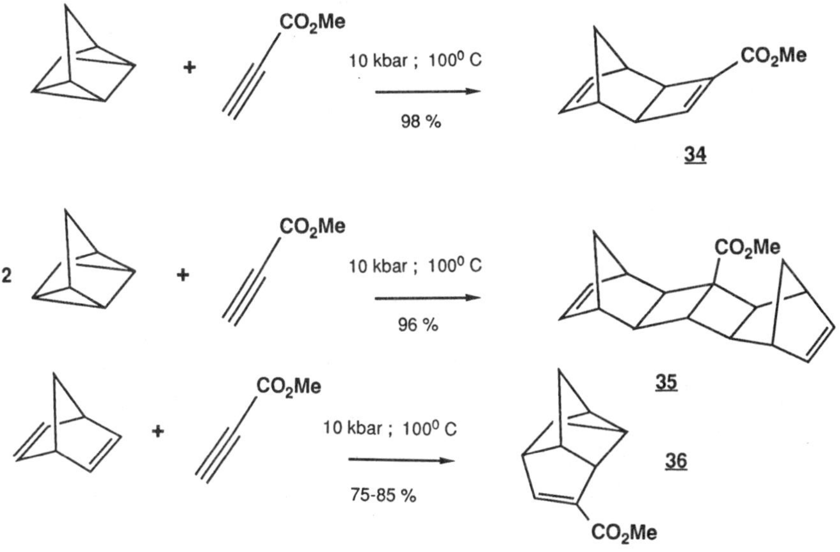

CO2Me
10 kbar ; 100° C
98 %
CO2Me
34
2
CO2Me
10 kbar ; 100° C
96 %
CO2Me
35
CO2Me
10 kbar ; 100° C
75-85 %
36
CO2Me

Scheme 55

The high-pressure Diels–Alder reaction of 1-methoxy-1,3-butadiene and aldehydes has been studied extensively and found to be useful for the preparation of 5,6-dihydro-2*H*-pyrans.[124]

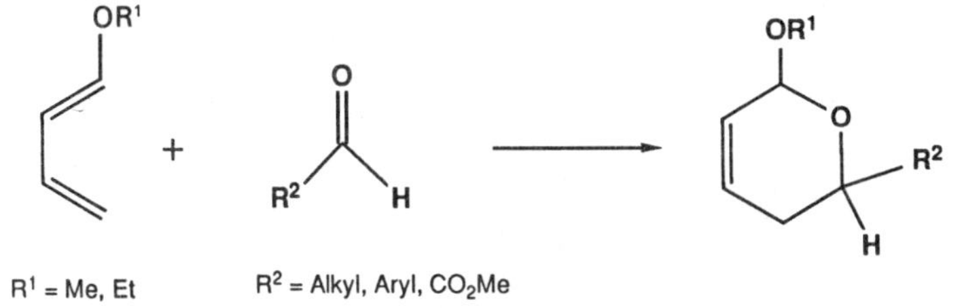

OR1
O
R2
H
OR1
O
R2
H
R1 = Me, Et
R2 = Alkyl, Aryl, CO2Me

Scheme 56

Extension of this approach to racemic N-protected α-aminoaldehydes permits entry to 7-amino functionalized 5,6-dihydro-2*H*-pyran compounds such as **37a–c**, which could serve as precursors for syntheses of complex monosaccharides[125] (e.g., aminoglycoside antibiotics,[126] purpurosamines). These hetero-Diels–Alder reactions require a lanthanide catalyst and a minimum pressure of 10 kbar.

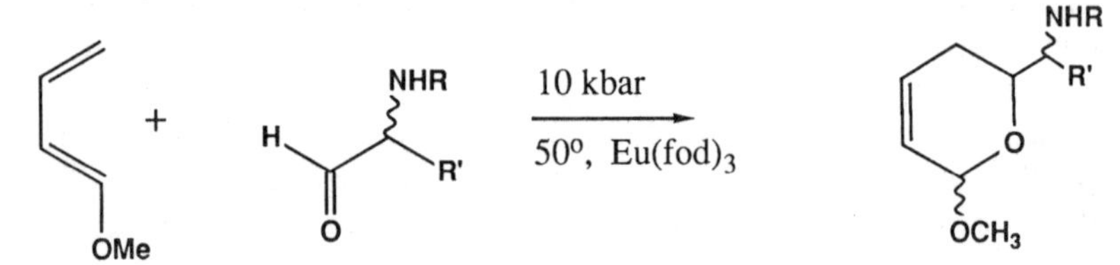

37 a-c

37a: R = Boc R′ = Me 27%

37b: R = Cbz R′ = Me 41%

37c: R = Tos R′ = CH$_2$O 67%

Scheme 57

Steric hindrance in the Boc-protected aldehydes is apparently responsible for the lower yields incurred in these reactions. Catalytic Pr(fod)$_3$ or Yb(fod)$_3$ may be substituted for Eu(fod)$_3$ to the same effect.[127]

A synthetic application involving this reaction type is the key step in the preparation of optically pure 4-deoxyheptosides, **38a,b**, as outlined in Scheme 58.[128]

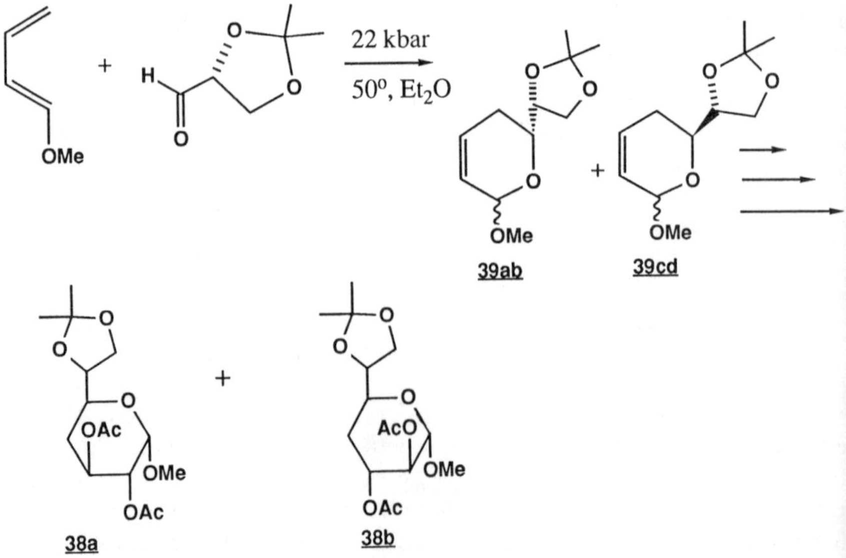

Scheme 58

The initial Diels–Alder reaction (22 kbar, 80%) gave a diastereomeric mixture of four products **39a–d** in a ratio of 66:16:13:05. Column chromatography was suitable to separate the diastereomers into two major fractions: **39a,b** and **39c,d**; fraction **39a,b** was employed as the starting material. Yields in the final steps were not reported so the overall efficiency of the five step sequence remains undisclosed.[129]

The first example of complete asymmetric induction in a noncatalyzed Diels–Alder has been reported by Jurczak and co-workers.[130] Reaction of 1-methoxybuta-1,3-diene and aldehyde **40** in diethyl ether at 20 kbar and 53°C for 20 h affords dihydropyran **41** as a single product in 72% yield.

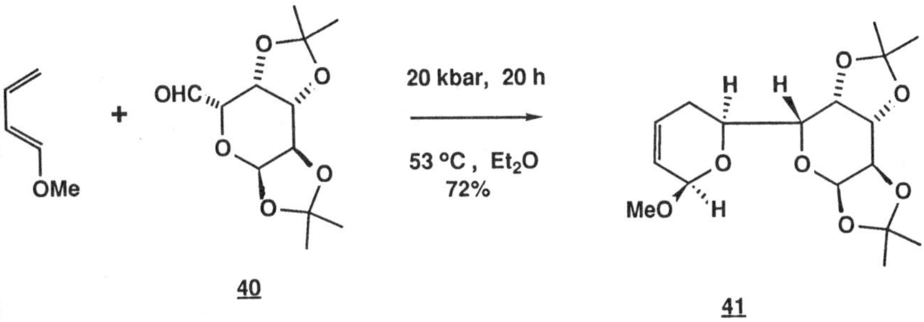

Scheme 59

Further work with sugar-derived aldehydes **42** and **43** provides similar diastereoselective excesses of up to 83%, although chemical yields of cycloadducts diminish considerably (**42**: 55%; **43**: 33%).

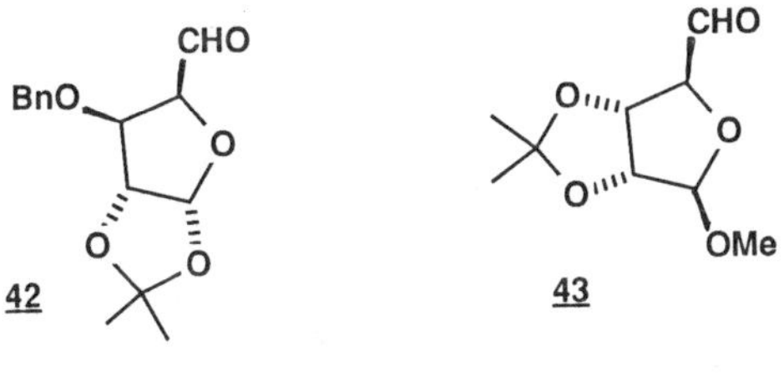

Scheme 60

Diastereoselectivity in the Diels–Alder reaction of enamino ketone **44** and ethyl vinyl ether is influenced significantly by pressure.[131] Highest selectivity for cis adduct **45** over trans **46** is obtained at low temperature and high pressure. For example, the ratio of **45** to **46** at 90°C and 1 bar is 1.67:1, but increases to 13.6:1 at 0.5°C and 6 kbar. The products are of interest because they can be readily converted to 3-amino sugars.[132]

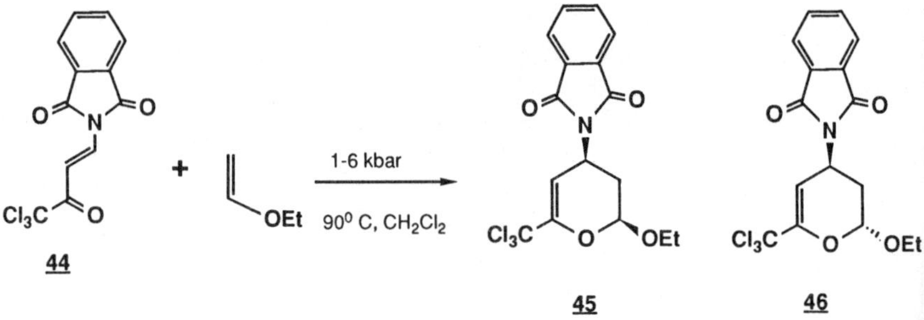

Scheme 61

5. Natural Product Synthesis

Several successful reports of the use of high pressure to effect key Diels–Alder reactions in the synthesis of natural products have been published in recent years. Synthetic targets constructed with the aid of high-pressure Diels–Alder reactions include cantharadin,[133] aklavinone,[134] massoilactone,[135] (+)-jatropholones A and B,[136] and ambreinolide.[137] A concise synopsis emphasizing the use of high-pressure methodology in these syntheses follows.

A preparative account of Dauben's succinct synthesis of cantharidin, **47**, first reported in 1980,[133] has appeared.[138] Optimum preparative scale conditions for the Diels–Alder reaction of furan and dihydrothiophene anhydride were achieved with excess furan, thereby allowing efficient conversion at low temperature and pressures suitable for large-scale reactions (7 kbar). This principal reaction can be carried out on a 10–15 g scale at 20°C in 24 h to 95% conversion. Development of lower pressure conditions is significant because of technical difficulties associated with scaling up higher pressure apparatus to conduct reactions at pressures greater than 7 kbar.

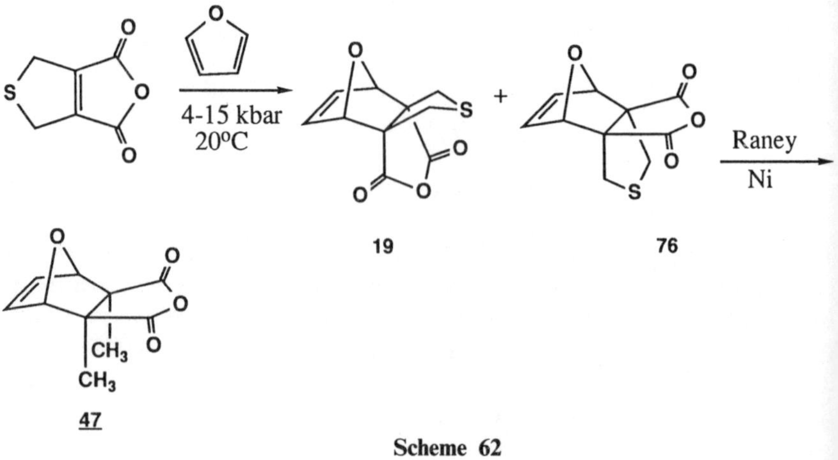

Scheme 62

An important step in Li and Wu's synthesis of aklavinone **48** is the Diels–Alder reaction of benzoquinone and diene ester **49** to afford enedione **50** in good yield.[134] Thermal attempts to conduct this reaction resulted in low yields.

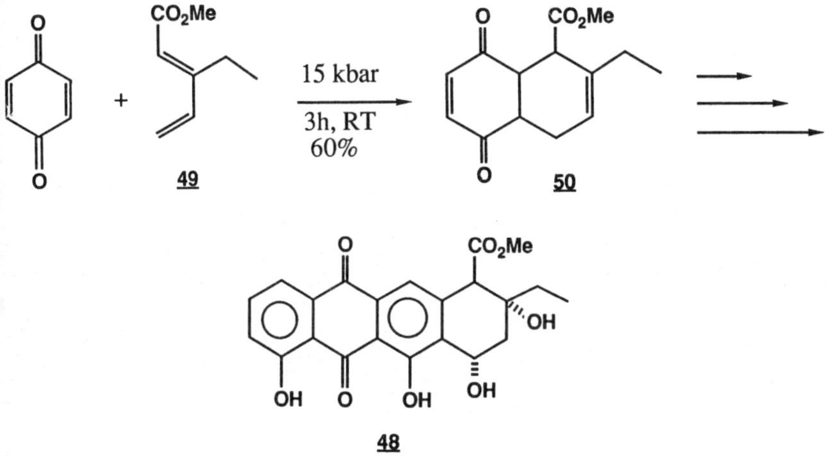

Scheme 63

The high-pressure reaction of 1-methoxybuta-1,3-diene and *n*-hexanal provides the key transformation, albeit in low yield (28%), for a synthesis of massoilactone **51**, outlined in Scheme 64.[135]

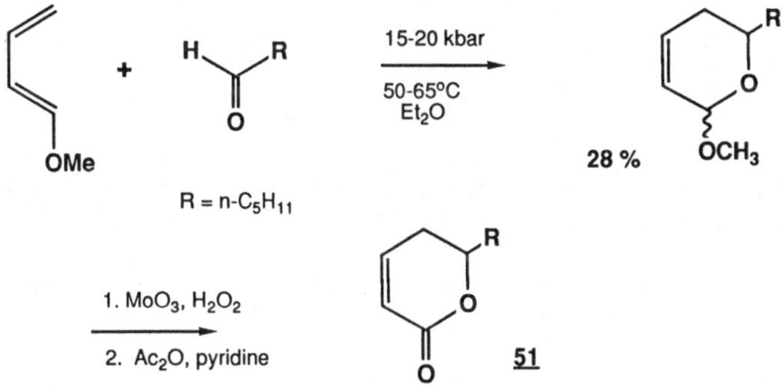

Scheme 64

The first total syntheses of the epimeric (+)-jatropholones A and B (**52**; **53**) have been achieved recently by Smith and co-workers.[136] The central transformation in this interesting account hinges on the Diels–Alder reaction of fused furan **54** and enone **55**. The researchers resorted to the use of high pressure because of the thermal instability of furan **54**.

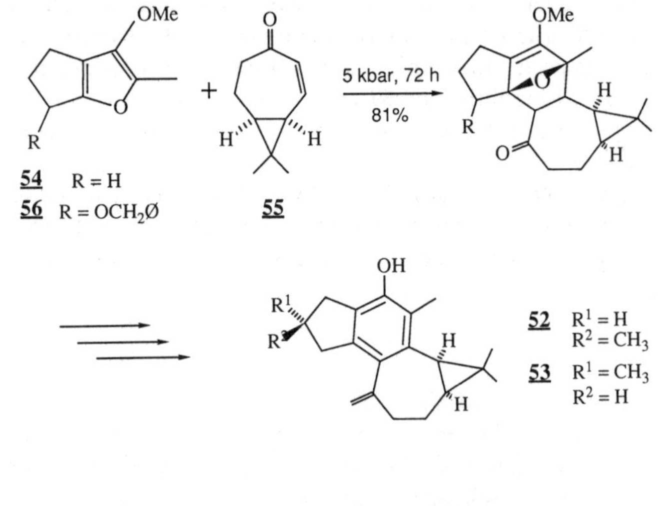

Scheme 65

Noteworthy is the fact that the analogous Diels–Alder reaction of furan **56** and enone **55** results in only 2.4% adduct formation under identical pressure conditions in over twice the reaction time (5 days). The electron-releasing and bulky benzoxy group in **56** is apparently responsible for the contrasting results and serves as a reminder of the sensitivity of high-pressure reactions to changes in structural features.

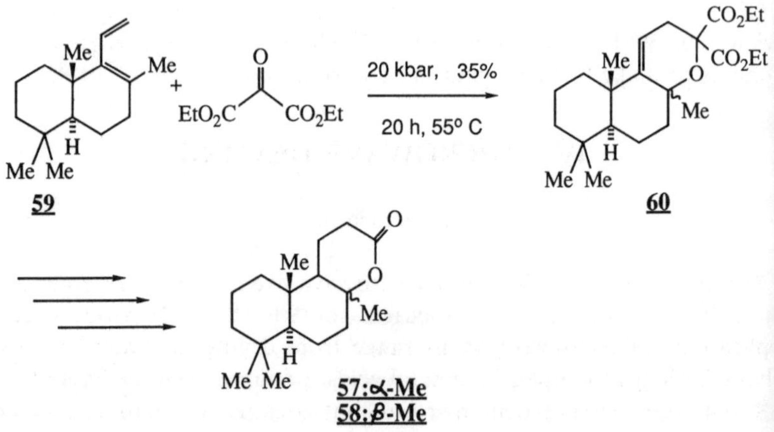

Scheme 66

A hetero-Diels–Alder reaction is the focal point in the synthesis of (±)-ambreinolide **57** and (±)-epiambreinolide, **58**.[137] (±)-Ambreinolide is a degradation product of ambrein[139] and is used in perfume production. Reaction of octahydronaphthalene **59** and diethyl mesoxalate at 55°C for 20 h at 20 kbar provides a diastereomeric mixture (65:35) of cycloadduct **60** (35%).

Hydrogenation of **60** (90%) is followed by diastereomer separation, but final degradation of the diester to the requisite lactone moiety in (±)-ambreinolide or (±)-epiambreinolide proceeds in low yield (20 and 18%, respectively).

C. Conclusion

Over 20 years have passed since le Noble's pioneering review of this technique, in which time this area of investigation has undergone tremendous growth. One hallmark of high-pressure methodology has been its ability to effect efficiently chemical transformations that proceed poorly under conventional conditions. In some cases reactions succeed only under high-pressure conditions. In addition, many high-pressure reactions may be conducted at lower temperatures, although reaction times can be long.

Despite potential drawbacks such as polymerization, difficult to predict solvent and temperature influences, as well as subtle substrate structural effects that can retard success of a given reaction, the method remains very attractive. This area of investigation continues to provide chemists with useful alternative approaches for reactions that are problematic, providing them a reliable tool for overcoming synthetic challenges. Moreover, recent work shows considerable progress being made in the area of asymmetric induction.

The use of high pressure in organic synthesis, described by one expert reviewer in 1981 as "in its infancy,"[88h] continues to expand rapidly and promises to make further significant contributions.

IV. MICROWAVE HEATING

A. Introduction

The first reports on the use of microwave ovens as thermal sources for conducting organic reactions appeared in 1986.[140,141] Previous scientific applications of microwave ovens range from drying glassware[142] to dissolving biological samples in mineral acids for subsequent analysis.[143]

Recent research has demonstrated that organic reactions can be conducted safely in commercial microwave ovens with reductions in reaction times of up to three orders of magnitude.[140,141] This time-saving ability is

the principal advantage of microwave heating. Further advantages inheren
in using this thermal source include the absence of significant temperature
gradients within the sample and very short response times.[143b] Moreover
microwave heating allows many solvents to play a direct role in the therma
process through their ability to couple with microwave irradiation, dis
tinguishing this heating mode from conventional ones.[144]

The mechanism of microwave heating is essentially that of dielectri
heating.[145] A simple model of an ac capacitor can be used to represent the
electric field vector of microwave irradiation: consider an array of molecule
situated between the plates of such a capacitor. The dipole moments of these
molecules (permanent and/or induced) are responsible for the molecule
experiencing significant torque as these dipoles align themselves with th
external electric field. As the polarity of the capacitor alternates, molecula
rotation is augmented as a result of attempted dipole realignment. The
energy absorbed in this process consequently increases the temperature o
the array. Microwave irradiation frequencies (2450 MHz in most commer
cial ovens) are of such magnitude to allow this electric field coupling t
occur. The ability of a substance to couple with microwave irradiation i
given by its molar polarizability, as derived from the Debye equation.[146]

Be advised that proper attention to safety precautions is necessary whe
conducting microwave experiments. A violent explosion was reported, fo
example, during attempts to oxidize toluene with potassium permanga
nate.[141] Our experience recommends that microwave reaction vessels b
housed in protective containers (constructed, for example, from Corian[14C]
and all reactions carried out in a functioning fume hood with protective sas
or shield.

B. Solvent Studies

Systematic solvent studies[147] have disclosed the existence of a qualitativ
relationship between the dielectric constant of a solvent and its ability t
couple with microwave irradiation.[140] A useful rule has emerged from thes
studies: the larger the dielectric constant, the greater the degree of couplin
hence the faster the temperature acquisition. Selected examples of dramat
differences in the ability of a solvent to couple with microwave irradiatio
are shown in Figure 1. Temperature data were obtained by using seale
capillaries charged with compounds of known melting point, as describe
elsewhere.[140]

Vermiculite has been found to be a convenient material for conductir
microwave reactions. The reaction vessels used in our laboratories[140] (seal
tubes or Ace Glass pressure tubes) are buried in a bed of vermiculi
contained in a rectangular box constructed of corian, a commercial

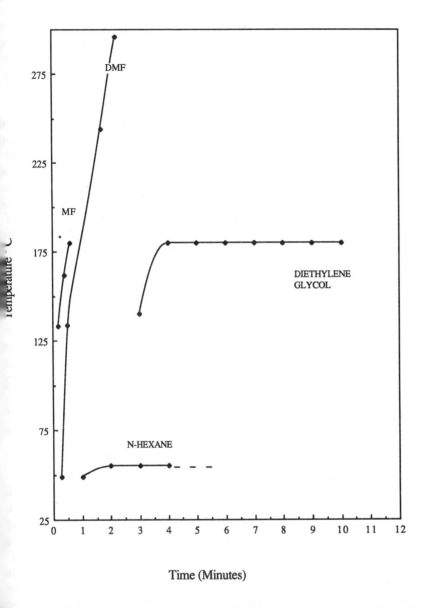

Time (Minutes)

Figure 1. Temperature approximations of selected solvents as a function of microwave irradiation time in the absence of insulation. An asterisk (*) denotes sealed tube explosion at indicated times. Dielectric constants (in Debye units): *N*-methylformamide (MF), 182.4; *N,N*-dimethylformamide (DMF), 36.7; diethylene glycol, 31.7; *n*-hexane, 1.9.

available heat-resistant polymer.[148] The contents of the reaction are
absorbed by vermiculite in the event of reaction vessel explosion, hence
vermiculite plays an important safety role in this regard.

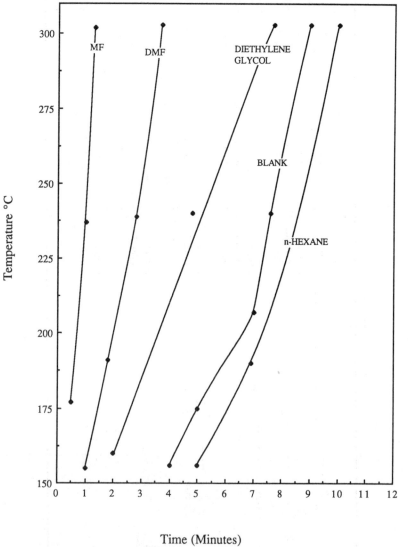

Time (Minutes)

Figure 2.   Temperature approximations of selected solvents as a function of
microwave irradiation time in vermiculite insulation. The blank refers to
data obtained testing an empty sealed tube. Dielectric constants (in Debye
units): *N*-methylformamide (MF), 182.4; *N,N*-dimethylformamide (DMF),
36.7; diethylene glycol, 31.7; *n*-hexane, 1.9.

Table 9.

Reaction	Conditions		
	Microwave	Literature	Control
1.	3 min, p-xylene, 92%, 160° < 187°C	10 min,[149] p-xylene, 90%, 138°C	10 min, p-xylene, 90%, 138°C
2.	10 min, p-xylene, 87%, 325° < 361°C	72 h,[150] dioxane, 90%, 101°C	4 h, p-xylene, 67%, 138°C
3.	12 min, neat, 55%, 325° < 361°C	5 h,[151] neat, 67%, 150°C	5 h, neat, 81%, 150°C
4.	10 min, neat, 66%, 325° < 361°C	4 h,[152] neat, 95%, 100°C	4 h, neat, 68%, 100°C
5.	15 min, neat, 25%, 400° < 425°C	72 h,[153] neat, 20%, 200°C	a. 3 days, neat,[154] 89%, 195°C b. 2 h, neat,[154] 75%, 195°C

157

Vermiculite consists of hydrous silicates of iron, aluminum, and magnesium. As a result, its water of hydration couples with microwave irradiation. Figure 2 shows irradiation time and temperature data for the solvents of Figure 1 when packed in vermiculite. A comparison of Figures 1 and 2 clearly shows that the use of vermiculite permits the heating of solvents unable to significantly couple with microwave irradiation.

C.   Reactions Conducted with Microwave Heating

1.   Diels–Alder, Claisen, and Ene Reactions

Microwave heating greatly reduces the time necessary to complete Diels–Alder reactions as illustrated in Table 9.[140] Reactions were conducted on a 1–2 mmol scale with vermiculite insulation. Control reactions and literature results are provided for comparision. Target temperatures of several hundred degrees are reached very rapidly, demonstrating the efficiency of this mode of heating. Isolated microwave yields are comparable to those in the literature and control runs. Of particular significance is the result of temperature-sensitive furan to readily provide oxabicyclic[2.2.1]heptadiene derivatives under these high-temperature conditions.

Table 10 presents comparative data from Claisen and ene reactions conducted in a manner similar to the Diels–Alder reactions.[140] Again a significant reduction in the time necessary to complete the reaction under microwave conditions is observed. Further reduction in reaction time is gained when N,N-dimethylformamide (DMF) or N-methylformamide (MF) is employed as solvent. The large dielectric constants of these solvents (cf. Section IV, B) promote efficient microwave coupling and allow substrates to acquire high temperatures extremely rapidly.

Recent work on the application of microwave heating to related reactions has led to optimization of the reaction of 1,4-cyclohexadiene and dimethylacetylene dicarboxylate to afford functionalized tricyclics **62**.[159] This process is a tandem ene/intramolecular Diels–Alder reaction, originally studied by Alder and Bong in 1952, but since neglected, presumably because of low yield (< 40%).[158] With microwave heating, this transformation can be achieved in up to 1 g scale with isolated yields of > 80% in less than 10 min.

The ene:enophile ratio significantly influences the outcome of the reaction. When the ratio is 1:1, the intermolecular Diels–Alder reaction of intermediate **61** and a second molecule of enophile predominates. This competing process is suppressed by use of excess diene. A 15:1 ene : enophile ratio leads to high yields of **62**, with either microwave or conventional heating. Control experiments require 20 h at > 220°C to complete, but

Table 10.

Reaction	Conditions		
	Microwave	Literature	Control
6.	a. 10 min, neat 21%, 325° < 361°C b. 6 min, DMF, 92%, 325° < 361°C	6 h,[155] neat, 85%, 220°C	6 min, neat, 17%, 320°C
7.	a. 12 min, neat 71%, 370° < 400°C b. 5 min, DMF, 72%, 300° < 315°C c. 90 sec, N-methylformamide, 87%, 276° < 300°C	85 min,[155] neat, 85%, 240°C	a. 45 min, neat, 71%, 265°C b. 12 min, neat, 92%, 320°C
8.	15 min, neat, 62%, 400° < 425°C	12 h,[156] neat, 85%, 180°C	12 h, neat, 60%, 180°C

159

conventional controls at 350–375°C at comparable times do not attain the microwave yields (Table 11, entries 2 and 5).

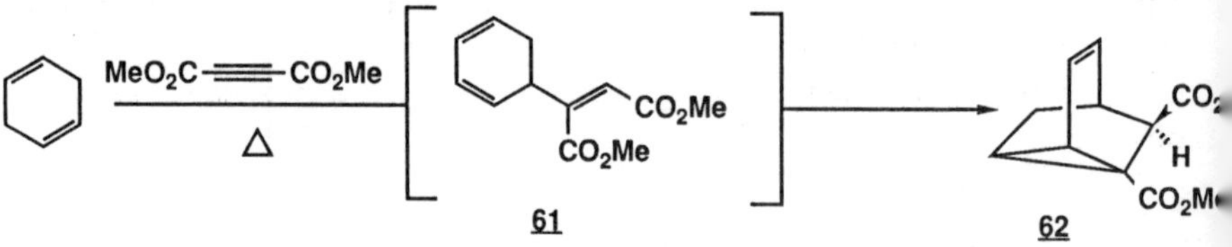

Table 11.

Thermal conditions	Ene : enophile ratio	Time	Temperature (°C)	% Yield
1. Literature[158]	1.1 : 1	20 h	180 < 190	40
2. Microwave	15 : 1	6 min	325 < 361	82
3. Microwave	30 : 1	6 min	317 < 325	87
4. Control	15 : 1	20 h	220	82
5. Control	15 : 1	15 min[a]	360	51

[a] As a safety precaution, microwave reactions are generally allowed to stand several minutes before the reaction vessel is removed from the vermiculite bed. Therefore, conventional controls were conducted at 15 min.

Extension of this approach to monosubstituted acetylene compounds affords analogous tricyclic adducts **63a,b** (Table 12), as well as polymer.[159]

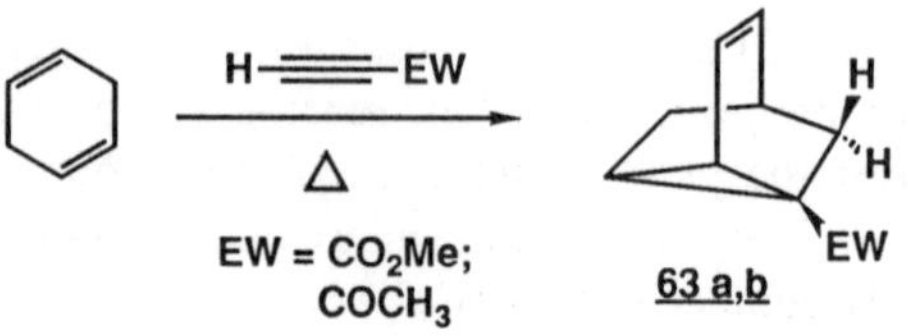

Table 12.

Thermal conditions	Time	Ene : enophile ratio	Temperature (°C)	% Yield
EW = CO_2CH_3 **(63a)**				
1. Microwave	8 min	Neat (10 : 1)	361 < 400°C	31
2. Control	21 h	Neat (10 : 1)	230 < 235°C	0
3. Control	3 h	Neat (10 : 1)	330° °C	30
EW = $COCH_3$ **(63b)**				
1. Microwave	7 min	Neat (10 : 1)	350 < 370°C	32
2. Control	3.5 h	Neat (10 : 1)	335 < 340°C	36

These results are significant because of the regioselectivity observed. Ene reactions of simple alkenes and unsymmetrical enophiles are known to produce isomeric mixtures in which the regioisomer resulting from alkene attack at the β-carbon of the enophile predominates.[160] Tricyclics **63a,b** clearly result from bond-formation at the α-carbon of the enophile[161] (Scheme 67). No evidence for the formation of β-regioisomers **64a,b** is observed.

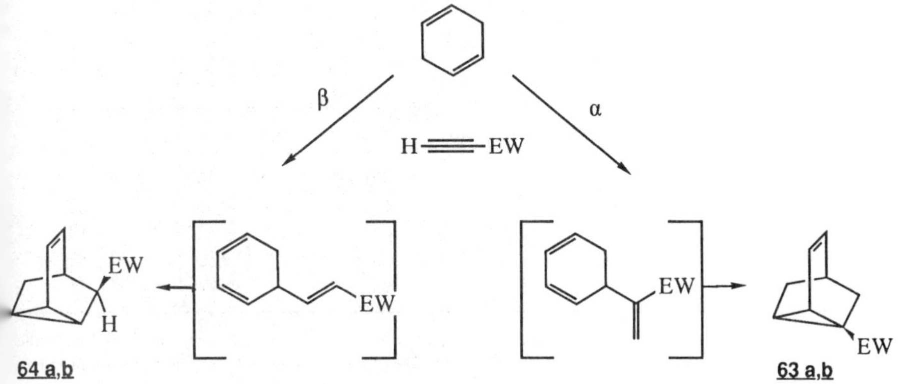

Scheme 67

In contrast to the dimethylacetylene dicarboxylate reactions, monosubstituted acetylene enophiles require higher temperatures (> 320°C) to react in reasonable time, as indicated by the failure of control reactions to produce adduct **63a**, even after 21 h at 235°C. Addition of commercial zinc chloride (20 mol%; based on enophile) allows the reactions to be conducted at 220°C in 4 h and increases the yield under conventional conditions to 68–77%. Yields remain low, however, under microwave conditions, presumably because of the short contact time with the catalyst.[159]

Tricyclics **63a,b** possess synthetic potential because they provide rapid entry into functionalized [2.2.2] bicyclic ring systems on opening of the cyclopropane ring.[162]

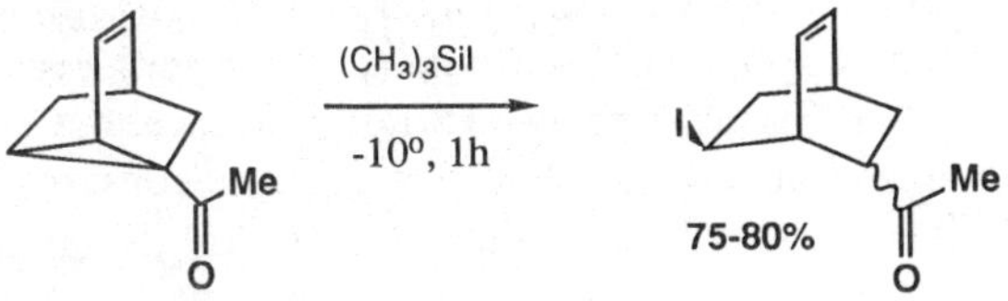

Scheme 68

A pilot study of the stereochemical influences of microwave heating on an intramolecular Diels–Alder reaction has been carried out.[157a,163] Microwave irradiation of trienal **65** in *p*-xylene (vermiculite insulated) for 8 min provides a mixture of stereoisomeric hydronaphthalenes **66a–d** shown in Scheme 69.

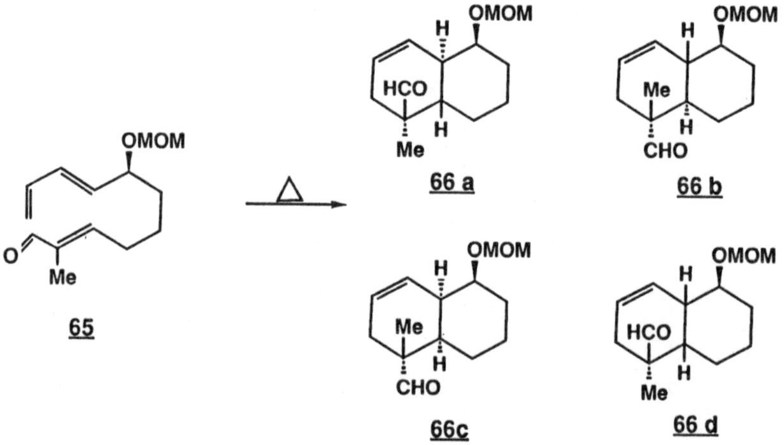

Thermal conditions	Solvent	Time	Temperature (°C)	Product ratio 66a:66b:66c: 66d
1. Literature[164]	Toluene	24 h	155	63:21:16:0
2. Microwave	*p*-Xylene	8 min	237 < 275	62:21:17:0

Scheme 69

The ratio of stereoisomers **66a–d** is essentially the same under either set of conditions. This indicates that microwave heating is influencing the reaction principally by thermal effects (presumably through coupling with water of hydration in the vermiculite and subsequent heat transfer). That is, either no direct coupling of microwave irradiation and the solvent and/or substrate is occurring, or such coupling effects are insignificant with regard to relative transition state energies and hence do not influence the stereochemical outcome of the reaction. The fact that the dielectric constant of *p*-xylene is small (2.3 Debye units) indicates that this solvent is unable to couple significantly with microwave irradiation, consistent with this interpretation.[140]

Table 13.

$$R\text{–}X \xrightarrow[\text{acetone, } \Delta]{\text{NaI}} R\text{–}I$$

$$X = Cl, Br$$

Thermal conditions	R	X	% Power	Time	Temperature (°C)	% Yield
1. Microwave	$n\text{-}C_6H_{13}$	Br	100	4 min	236 < 275	68
2. Microwave	$n\text{-}C_6H_{13}$	Br	15	15 min	175 < 187	90
3. Control	$n\text{-}C_6H_{13}$	Br	—	30 min	56	95
4. Microwave	$n\text{-}C_{12}H_{25}$	Cl	25	15 min	325 < 361	93
5. Control	$n\text{-}C_{12}H_{25}$	Cl	—	1 h	56	17[a]
6. Microwave	MeO-⟨C₆H₄⟩-(CH₂)₄	Br	15	8 min	175 < 187	76
7. Control	MeO-⟨C₆H₄⟩-(CH₂)₄	Br	—	1 h	56	92
8. Microwave	$AcO\text{-}(CH_2)_4$	Cl	15	15 min	175 < 187	>95[b]
9. Control	$AcO\text{-}(CH_2)_4$	Cl	—	1 h	56	26[b]

[a] NMR estimation.
[b] Product contains 10–13% 1,4-diiodobutane by GLC.

163

2. Nucleophilic Substitution and Dehydration

Microwave heating is capable of promoting nucleophilic substitution reactions. Two reaction types have been investigated to date: Finkelstein reactions afford alkyl iodides from alkyl chlorides and bromides[157a]; reaction of 4-cyanophenoxide ion and benzyl chloride gives aryl ether **67**.[141]

Table 13 compares microwave and conventional results for the preparation of alkyl iodides under standard Finkelstein conditions (1.5 equivalents sodium iodide, dry acetone).[165] As anticipated, alkyl bromides are converted more readily than alkyl chlorides under conventional conditions. In contrast, under microwave conditions the Finkelstein reactions appear independent of the nature of halide. This difference is attributed to the superheating conditions generated under microwave conditions, which presumably leads to rapid equilibration of reaction species. For alkyl chlorides, precipitation of the poorly soluble sodium chloride provides a driving force to enhance the forward reaction.

Reductions in the time necessary to complete formation of aryl ether **67** of up to three orders of magnitude have been reported. In addition, the influence of pressure on this reaction was studied. Taking the reaction to 65% completion in three different-sized vessels provided kinetic data indicating reaction rate increases to be proportional to increasing pressure.[141]

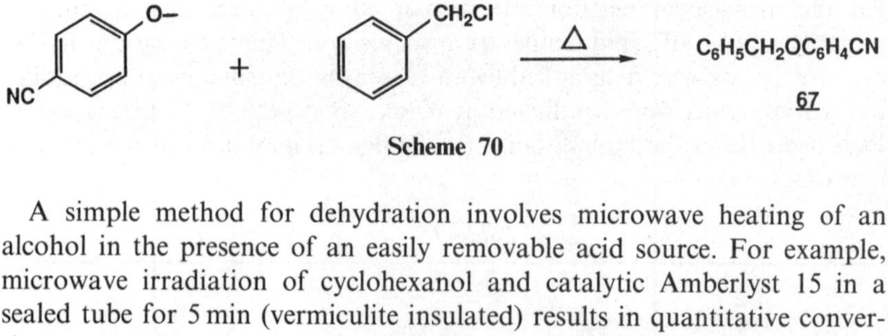

Scheme 70

A simple method for dehydration involves microwave heating of an alcohol in the presence of an easily removable acid source. For example, microwave irradiation of cyclohexanol and catalytic Amberlyst 15 in a sealed tube for 5 min (vermiculite insulated) results in quantitative conversion to cyclohexene. Control experiments in the absence of the acid resin afford no cyclohexene, hence a purely pyrolytic pathway is excluded.[157a] Table 14 presents results from the dehydration of 2-phenyl-2-propanol.[166] Significant is the ability to afford principally either α-methyl styrene **68** or indane **69** as a function of irradiation time.

3. Miscellaneous Reactions

Single examples of the following reaction types have been examined under microwave conditions: hydrolysis of benzamide and methyl benzoate, oxidation of toluene,[141a] oxime formation, and preparation of naphoxyacetic acid from phenol and chloroacetic acid.[141b]

Table 14.

| | | | | % Yield | |
Thermal conditions	Solvent	Time	Temperature (°C)	68	69
1. Microwave	Neat	2 min	276 < 299	98[a]	—
2. Microwave	Neat	6 min	325 < 361	—	88
3. Control	Neat	48 h	RT	93[a]	—
4. Control	Neat	19 h	120–130	—	80

[a] Contains 12–14% open chain intermediate, 2,4-diphenyl-4-methyl-2-pentene, by NMR.

A study of benzoic acid esterification has also been conducted as presented in Table 15.[141] Reductions in time necessary to complete these classic reactions range from 8- to 96-fold; greatest reductions occur with methanol. Esterification with n-pentanol at the same temperature provided only a slight decrease in the reaction time under microwave conditions. This indicates that the increase in reaction rates observed with microwave heating is primarily a result of rapid temperature acquisition. The temperatures in the microwave runs were determined by an IR sensor immediately after irradiation; control runs were conducted at reflux. As expected, control reaction times decrease as the boiling point (i.e., reflux temperature) of the alcohol increases.

Table 15.

$$R = CH_3;$$
$$n\text{-}C_3H_7; n\text{-}C_4H_9$$

Thermal conditions	R	Time	Temperature (°C)	% Yield
1. Microwave	CH_3	5 min	134	76
2. Control	CH_3	8 h	65	74
3. Microwave	$n\text{-}C_3H_7$	6 min	135	79
4. Control	$n\text{-}C_3H_7$	4 h	97	78
5. Microwave	$n\text{-}C_4H_9$	7.5 min	135	79
6. Control	$n\text{-}C_4H_9$	1 h	117	82

D. Conclusion

Microwave heating in organic synthesis is a new area of investigation. Although some progress has been made, much remains to be explored and accomplished. Reductions in reaction time clearly offer advantages to the bench chemist and this methodology permits very efficient acquisition of high reaction temperatures. Unlike flash vacuum pyrolysis, this technique permits high temperature study of *inter*molecular reactions. The direct role a solvent is able to play in the thermal process is unique to microwave heating and offers the potential for unexpected chemical behavior.

Judging from the magnitude of response from the scientific community, application of microwave heating to organic synthesis is a field certain to experience significant future cultivation.[167]

VI. SUMMARY

The use of nonconventional reaction conditions in organic synthesis has led to the outgrowth of significant new developments in the field. The principal advantages of these methods are (1) reductions in time necessary to complete organic reactions, (2) milder reaction conditions, (3) often increased reaction yields (especially with ultrasound), and (4) the accomplishment of synthetic transformations that fail using traditional methods.

Moreover, novel synthetic outcomes have been observed when employing ultrasound or high-pressure conditions to certain organic reactions, providing additional incentive to explore new dimensions of these methodologies. Future investigation will reveal if microwave heating is capable of similarly influencing organic synthesis.

REFERENCES

1. Attarwala, S. T. Application of High Pressure to Organic Synthesis. Dissertation, Polytechnic Institute of New York, 1984.
2. (a) Giguere, R. J.; Bray, T. L.; Duncan, S. M.; Majetich, G. *Tetrahedron Lett.* **1986**, *27*, 4945. (b) Gedye, R.; Smith, F.; Westaway, K.; Humera, A.; Baldisera, L.; Laberge, L.; Rousell, J. *Tetrahedron Lett.* **1986**, *27*, 279.
3. A DIALOG search for the singular term "flash vacuum pyrolysis" identified 88 articles from 1982 to 1986. For concise reviews of FVT see Schiess, P.; Rutschmann, S. *Chimia* **1985**, *39*, 213; Wiersum, U. E. *Aldrichim. Acta* **1984**, *17*, 31, and references therein.
4. Richards, W. T.; Loomis, A. L. *J. Am. Chem. Soc.* **1927**, *49*, 3087.
5. Porter, C. W.; Young, L. *J. Am. Chem. Soc.* **1938**, *60*, 1497.
6. Renaud, P. *Bull. Soc. Chim. Fr.* **1950**, 1044.

7. For reviews on the use of ultrasound in organic synthesis see (a) Boudjouk, P. In *High Energy Processes in Organometallic Chemistry*, Suslick, K. S., ed., ACS Symposium Series 333, American Chemical Society, Washington, DC, 1987, pp. 209–222. (b) Suslick, K. S. In *Modern Synthetic Methods*, Scheffold, R., ed., Springer-Verlag, Berlin, 1986, pp. 1–60. (c) Suslick, K. S. *Adv. Organomet. Chem.* **1986**, *25*, 73. (d) Boudjouk, P. *J. Chem. Ed.* **1986**, *63*, 427. (e) Toma, S.; Kaliska, V. *Chem. Listy* **1985**, *79*, 578. (f) Kitazume, T.; Ishikawa, N. *Senryo Yakuhin* **1985**, *30*(6), 175.

8. Sehgal, C.; Sutherland, R. G.; Verrall, R. E. *J. Phys. Chem.* **1980**, *84*, 388.

9. Flynn, H. G. In *Physical Acoustics*, Mason, W. P., ed., Academic Press, New York, 1964, Vol. 1B.

10. (a) Suslick, K. S.; Hammerton, D. A.; Cline, R. E., Jr. *J. Am. Chem. Soc.* **1986**, *109*, 5641. (b) Suslick, K. S., Cline, R. E., Jr.; Hammerton, D. A. *Ultrason. Symp. Proc.* **1985**, *2*, 1116.

11. (a) Mason, T. J. *Lab. Pract.* **1984**, *33*, 13. (b) Mason, T. J.; Lorimer, J. P.; Mistry, B. P. *Tetrahedron Lett.* **1983**, *24*, 4371. (c) Mason, T. J.; Lorimer, J. P.; Mistry, B. P. *Tetrahedron Lett.* **1982**, *23*, 5363. (d) Lorimer, J. P.; Mason, T. J. *J. Chem. Soc. Chem. Commun.* **1980**, 1135.

12. Kristol, D. S.; Klotz, H.; Parker, R. C. *Tetrahedron Lett.* **1981**, *22*, 907.

13. For ultrasonic investigations into two-phased hydrolysis of esters see Moon, S.; Duchin, L.; Cooney, J. V. *Tetrahedon Lett.* **1979**, *41*, 3917. See also ref. 64.

14. For a discussion of the effect of solvent vapor pressure on sonication of alkanes, see (a) Suslick, K. S.; Gawlenowski, J. J.; Schubert, P. F.; Wang, H. H. *J. Phys. Chem.* **1983**, *87*, 2299. (b) Couppis, E. C.; Klinzing, G. E. *AIChE J.* **1974**, *20*(3), 485.

15. (a) Margulis, M. A. *Russian J. Phys. Chem.* **1976** *50*(1), 1. (b) Flynn, H. G. *J. Acoust. Soc. Am.* **1975**, *58*, 1160 and references therein. (c) Plesset, M. S.; Prosperetti, A. *Annu. Rev. Fluid Mech.* **1977**, *9*, 145. (d) Prosperetti, A. *J. Acoust. Soc. Am.* **1975**, *57*, 810 and references therein.

16. Luche, J. L.; Petrier, C.; Germal, A. L.; Zikra, N. *J. Org. Chem.* **1982**, *47*, 3806.

17. Luche, J. L.; Petrier, P.; Lansard, J. P.; Greene, A. E. *J. Org. Chem.* **1983**, *48*, 3837.

18. (a) Petrier, C.; de Souza Barbosa, J. C.; Duprey, C.; Luche, J. L. *J. Org. Chem.* **1985**, *50*, 5761. (b) For a related study of conjugate addition to enals, see de Souza Barbosa, J. C.; Petrier, C.; Luche, J. L. *Tetrahedron Lett.* **1985**, *26*, 829.

19. Petrier, C.; Luche, J. L.; Dupuy, C. *Tetrahedron Lett.* **1984**, *25*, 3463.

20. Petrier, C.; Dupuy, C.; Luche, J. L. *Tetrahedron Lett.* **1986**, *27*, 3149.

21. For use of organocopper reagents in synthesis, see Posner, G. H. *Introduction to Synthesis Using Organocopper Reagents*, Wiley Interscience, New York, 1980.

22. Han, B. H.; Boudjouk, P. *J. Org. Chem.* **1982**, *47*, 5030.

23. Bose, A. K.; Gupta, K.; Manas, M. S. *J. Chem. Soc. Chem. Commun.* **1984**, 86.

24. Flitsch, W.; Russkamp P. *Liebigs Ann. Chem.* **1985**, 1398. Antitumor agent mitomycin C has recently been found to crosslink DNA: Tomasz, M.; Lipman, R.; Chowdray, D.; Pawlack, J.; Verdine, G.; Nakanishi, K. *Science* **1987**, *235*, 1204.

25. Iskikawa, N.; Koh, M. G.; Kitazume, T.; Choi, S. K. *J. Fluorine Chem.* **1984**, *24*, 419.

26. Petrier, C.; Luche, J. L. *J. Org. Chem.* **1985**, *50*, 912. For additional studies see Einhorn, C.; Luche, J. L. *J. Organomet. Chem.* **1987**, *322*(2), 177.

27. Molle, G.; Bauer, P. *J. Am. Chem. Soc.* **1982**, *104*, 3481.

28. Petrier, C.; Einhorn, J.; Luche, J. L. *Tetrahedron Lett.* **1985**, *26*, 1449. For examples of such selectivity with titanium reagents see Kostova, K.; Hesse, M. *Helv. Chim. Acta* **1984**, *67*, 1713 and Weidemann, B.; Seebach, D. *Angew. Chem. Int. Ed. Engl.* **1983**, *22*, 31.

29. Knochel, P.; Normant, J. F. *Tetrahedron Lett.* **1984**, *25*, 1475.

30. For reviews of this reaction, see (a) Hoffmann, H. M. R. *Angew. Chem. Int. Ed. Engl.* **1984**, *23*, 1. (b) Mann, J. *Tetrahedron* **1986**, *42*, 4611. (c) Noyori, R.; Hayakawa, Y. *Org. React.* **1983**, *29*, 163.

31. Joshi, N. N.; Hoffmann, H. M. R. *Tetrahedron Lett.* **1986**, *27*, 687.

32. Repic, O.; Vogt, S. *Tetrahedron Lett.* **1982**, *23*, 2729.

33. Yamashita, J.; Inoue, Y.; Kondo, T.; Hashimoto, H. *Bull. Chem. Soc. Jpn.* **1984**, *57*, 2335.

34. (a) Miyano, S.; Hida, H; Hashimoto, H. S. *Koygo Kaqku Zasshi* **1966**, *69*, 2134. (b) Hashimoto, H.; Hida, M.; Miyano, S. *J. Organomet. Chem.* **1967**, *10*, 518. (c) Miyano, S.; Hida, M.; Hashimoto, H. *J. Organomet. Chem.* **1968**, *12*, 263.

35. Regen, S. L.; Singh, A. *J. Org. Chem.* **1982**, *47*, 1587.

36. (a) Zeichmeister, L.; Wallcave, L. *J. Am. Chem. Soc.* **1955**, *77*, 2853. (b) Zeichmeister, L.; Currell, D. L. *J. Am. Chem. Soc.* **1958**, *80*, 205.

37. Prakash, S.; Pandey, J. D. *Tetrahedron Lett.* **1965**, *21*, 903.

38. Han, B. H.; Boudjouk, P. *Tetrahedron Lett.* **1981**, *22*, 2757.

39. Boudjouk, P.; Han, B. H. *Tetrahedron Lett.* **1981**, *22*, 3813.

40. Boudjouk, P.; Han, B. H.; Anderson, K. R. *J. Am. Chem. Soc.* **1982**, *104*, 4992.

41. (a) Masamune, S.; Murakami, S.; Tobita, H. *Organometallics* **1983**, *2*, 1464. (b) Boudjouk, P. *J. Chem. Ed.* **1986**, *63*, 427.

42. Luche, J. L.; Damiano, J. C. *J. Am. Chem. Soc.* **1980**, *102*, 7926.

43. Renaud, P. *Bull. Soc. Chim. Fr.* **1950**, 1054.

44. Pearce, P. J.; Richards, D. H.; Scilly, N. F. *J. Chem. Soc. Perkin Trans. I* **1972**, 1655.

45. Burkov, I.; Syndes, L. K.; Ubeda, D. C. N. *Acta Chem. Scand., Ser. B* **1987**, *B41*, 235.

46. de Souza-Barbosa, J. C.; Petrier, C.; Luche, J. L. *J. Org. Chem.* **1988**, *53*, 1212. For related studies, see ref. 66.

47. Petrier, C.; Gemal, A.; Luche, J. L. *Tetrahedron Lett.* **1982**, 3361.

48. (a) Einhorn, J.; Luche, J. L. *Tetrahedron Lett.* **1986**, *27*, 1791. (b) Einhorn, J.; Luche, J. L. *Tetrahedron Lett.* **1986**, *27*, 1793.

49. Luche, J. L.; Petrier, C.; Dupuy C. *Tetrahedron Lett.* **1984**, *25*, 753.

50. Sinisterra, J. V.; Fuentes, A.; Marinas, J. M. *J. Org. Chem.* **1987**, *52*, 3875.

51. Brown, H. C.; Racherla, U. S. *Tetrahedron Lett.* **1985**, *26*, 2187.

52. (a) Brown, H. C.; Racherla, U. S. *J. Org. Chem.* **1986**, *51*, 427. (b) Brown, H. C.; Racherla, U. S. *Tetrahedron Lett.* **1985**, *26*, 4311.

53. Liou, K. F.; Yang, P. H.; Lin, Y. T. *J. Organomet. Chem.* **1985**, *294*, 145.

54. (a) Kitazume, T.; Ishikawa, N. *J. Am. Chem. Soc.* **1985**, *107*, 5186. (b) Kitazume, T.; Ishikawa, H. *Chem. Lett.* **1981**, 1679.

55. Kitazume, T.; Ishikaua, N. *Chem. Lett.* **1982**, 137.

56. Kitazume, T.; Ishikaua, N. *Chem. Lett.* **1982**, 1453.

57. Enders, D. *Chemtech* **1981**, 504.

58. Solladie-Cavallo, A.; Farkhani, D.; Fritz, S.; Lazrak, T.; Suffert, J. *Tetrahedron Lett.* **1984**, *25*, 4117.

59. (a) Azuma, T.; Yanagida, S.; Sakurai, H. *Synth. Commun.* **1982**, *12*, 137. (b) For ultrasonic preparation and applications of anthracene magnesium, see Oppolzer, W.; Schneider, P. *Tetrahedron Lett.* **1984**, *25*, 3305.

60. Boudjouk, P.; Sooriyakumarin, R.; Han, B. H. *J. Org. Chem.* **1986**, *51*, 2818.

61. Han, B. H.; Boudjouk, P. *J. Org. Chem.* **1982**, *47*, 752.

62. Han, B. H.; Boudjouk, P. *Tetrahedron Lett.* **1982**, *23*, 1643.

63. Brown, H. C.; Krishnamurthy, S. *J. Org. Chem.* **1969**, *34*, 3918.

64. Boudjouk, P.; Han, B. H. *J. Catal.* **1983**, *79*, 489.

65. Han, B. H.; Boudjouk, P. *Organometallics* **1983**, *2*, 769.

66. Suslick, K. S.; Casadonte, D. J. *J. Am. Chem. Soc.* **1987**, *109*, 3459. For preparation of active metals (Zn, Cu, Ni) by sonochemical reduction of metal halides with lithium, see Boudjouk, P.; Thompson, D. P.; Ohrbom, W. H.; Han, B. H. *Organometallics* **1986**, *5*(6), 1257.

67. Ando, T.; Kawate, T.; Ichihara, J.; Hanafusa, T. *Chem. Lett.* **1984**, 725.
68. Ando, T.; Kawate, T.; Yamawaki, J.; Hanafusa, T. *Synthesis* **1983**, 637.
69. Priebe, H. *Acta Chem. Scand. Ser. B* **1984**, *38*, 895.
70. Ezquerra, J.; Alvarez-Builla, J. *J. Chem. Soc. Chem. Commun.* **1984**, 55.
71. Davidson, R. S.; Patel, A. M.; Safder, A.; Thornthwaite, D. *Tetrahedron Lett.* **1983**, *24*, 5907.
72. Fujita, T.; Watanabe, S.; Sakamoto, M.; Hashimoto, H. *Chem. Ind. (London)* **1986**, 427.
73. Mousseron, M.; Jacquier, R.; Christol, H. *Bull. Soc. Chim. Fr.* **1957**, 346.
74. Merrindey, J. C.; Trigo, G. G.; Sollhuber, M. M. *Tetrahedron Lett.* **1986**, *27*, 3285.
75. Yamawaki, J.; Sumi, S.; Ando, T.; Hanafusa, T. *Chem. Lett.* **1983**, 379.
76. Einhorn, J.; Luche, J. L. *J. Org. Chem.* **1987**, *52*, 4124.
77. For ultrasonic preparation of dimsylsodium see Sjoberg, K. *Tetrahedron Lett.* **1966**, 6383.
78. Hidehito, K.; Katsuyoshi, U.; Matsui, M. *Chem Express* **1987**, *2*, 169.
79. Hanafusa, T.; Ichihara, J.; Ashida, T. *Chem. Lett.* **1987**, 687.
80. Raucher, S.; Klein, P. *J. Org. Chem.* **1981**, *46*, 3558.
81. (a) Fry, A. J.; Herr, D. *Tetrahedron Lett.* **1978**, *20*, 1721. (b) Fry, A. J.; Ginsburg, G. S.; Parente, R. A. *J. Chem. Soc. Chem. Commun.* **1978**, 1040.
82. Kim, H. K.; Matyjaszewski, K. *J. Am. Chem. Soc.* **1988**, *110*, 3321.
83. Thompson, D. P.; Boudjouk, P. *J. Org. Chem* **1988**, *53*, 2109.
84. Einhorn, C.; Luche, J. L. *Carbohydr. Res.* **1986**, *155*, 258.
85. (a) Suslick, K. S.; Schubert, P. F.; Goodale, J. W. *J. Am. Chem. Soc.* **1981**, *103*, 7342. (b) Suslick, K. S.; Goodale, J. W.; Schubert, P. F.; Wang, H. H. *J. Am. Chem. Soc.* **1983**, *105*, 5781.
86. Suslick, K. S.; Johnson, R. E. *J. Am. Chem. Soc.* **1984**, *106*, 6856. For a review of the chemical effects of ultrasound on organometallic systems, see Suslick, K. S. In *High Energy Processes in Organometallic Chemistry*, Suslick, K. S., ed., ACS Symposium Series 333, American Chemical Society, Washington, DC, 1987, pp. 191–208.
87. The following consideration has been put forth to support this empirical observation: sonication of low vapor pressure solvents results in the formation of microbubbles that are nearly void. Collapse of these bubbles presumably produces more energetic shock waves than collapse of partially vapor-filled microbubbles that result from sonication of high vapor pressure solvents.[7d]
88. (a) Issacs, N. S.; George, A. N. *Chem. Br.* **1987**, 47. (b) Eldik, R. von; Jonas, J. *High Pressure Chemistry and Biochemistry*, D. Reidel, Dordrecht, 1987. (c) le Noble, W. J. *NATO ASI Ser., Ser. C.* **1987**, *197*, 295. (d) Appl, M. *Indian Chem. Eng.* **1987**, *29*, 7. (e) Jurczak, J. J. *Phys. Status Solidi [Sect.] B* **1986**, *139/140*. 709. (f) Attarwala, S. T. Application of High Pressure to Organic Synthesis. Dissertation, Polytechnic Institute of New York, 1984. University Microfilms International, No. 8420440, Ann Arbor. (g) Welzel, P. *Nachr. Chem. Techn. Lab.* **1983**, *31*, 148. (h) Matsumoto, K.; Uchida, T.; Acheson, R. M. *Heterocycles* **1981**, *16*, 1367. (i) le Noble, W. J.; Kelm, H. *Angew. Chem.* **1980**, *19*, 841, and leading references. (j) Gladysz, J. A. *Chemtech* **1979** 372, and references therein. (k) le Noble, W. J. *J. Chem. Ed.* **1967**, *44*, 729.
89. (a) Matsumoto, K.; Sera, A.; Uchida, T. *Synthesis* **1985**, 1. (b) Matsumoto, K.; Sera, A. *Synthesis* **1985**, 999.
90. (a) N. S. Isaacs, *Liquid Phase High Pressure Chemistry*, John Wiley, Chichester, 1981. (b) Isaacs, N. S. *Annu. Rep. Prog. Chem.* **1981**, *78*(B), 29. (c) H. Kelm, ed., *High Pressure Chemistry*, Proceedings of NATO Advanced Study Institute, Reidel, Amsterdam, 1978. (d) Asano, T., le Noble, W. J. *Chem. Rev.* **1978**, *78*, 407. (e) Hagan, A. P. *J. Chem. Ed.* **1978**, *55*, 620. Cf. also reference 88j and 88k.
91. le Noble, W. J. *Progr. Phys. Org. Chem.* **1967**, *5*, 207.

92. The pascal is the IUPAP unit of pressure: one pascal equals one newton/m^2; 1 MPa (megapascal) equals 0.01 kilobar.

93. For a discussion of increasing solvent viscosity enhancing reaction rates, cf. Firestone, R. A.; Vitak, M. A. *J. Org. Chem.* **1981**, *46*, 2160.

94. Dauben, W. G.; Gerdes, J. M.; Bunce, R. A. *J. Org. Chem.* **1984**, *49*, 4293.

95. (a) Pietraszkiewicz, M.; Salanki, P.; Jurczak, J. *J. Chem. Soc. Chem. Commun.* **1983**, 1184. (b) For a brief review of cryptand high pressure work, see Jurczak, J.; Ostaszewski, R.; Pietraszkiewicz, M.; Salanski, P. *J. Inclusion Phenom.* **1987**, *5*, 553.

96. Pietraszkiewicz, M.; Salanki, P.; Jurczak, J. *Heterocycles* **1985**, *23*, 547. For a review see Jurczak, J.; Pietraszkiewicz, M. *Topics Curr. Chem.* **1986**, *130*, 183.

97. Dauben, W. G., Gerdes, J. M.; Look, G. C. *J. Org. Chem.* **1986**, *51*, 4964.

98. (a) Noyori, R.; Murata, S.; Suzuki, M. *Tetrahedron* **1981**, *37*, 3899. (b) Tsunoda, T.; Suzuki, M.; Noyori, R. *Tetrahedron Lett.* **1980**, *21*, 1357.

99. Conjugate addition of methanol to enones under high pressure is known, see Scott, J. J.; Bower, K. R. *J. Am. Chem. Soc.* **1967**, *89*, 2682.

100. Dauben, W. G.; Gerdes, J. M.; Look, G. C. *Synthesis* **1986**, 532.

101. Midland, M. M.; McLoughlin, J. I. *J. Org. Chem* **1984**, *49*, 1317.

102. Midland, M. M.; Zderic, S. A. *J. Am. Chem. Soc.* **1982**, *104*, 525.

103. Midland, M. M.; Petre, J. E.; Zderic, S. A., Kazubski, A. *J. Am. Chem. Soc.* **1982**, *104*, 528.

104. d'Angelo, J.; Maddaluno, J. *J. Am. Chem. Soc.* **1986**, *108*, 8112.

105. Kinas, R.; Pankiewicz, K.; Stec, W. J.; Farmer, P. B.; Foster, A. B.; Jarmen, M. *J. Org. Chem.* **1977**, *42*, 1650.

106. Sera, A.; Takagi, K.; Katayama, H.; Yamada, H.; Matsumoto, K. *J. Org. Chem.* **1988**, *53*, 1157.

107. Aben, R. W. M.; Scheeren, H. W. *Tetrahedron Lett.* **1983**, *24*, 4613.

108. Aben, R. W. M.; Scheeren, H. W. *Tetrahedron Lett.* **1985**, *26*, 1889.

109. Bunce, R. A.; Schlecht, M. F.; Dauben, W. G.; Heathcock, C. H. *Tetrahedron Lett.* **1983**, *24*, 4943. Yamamoto, Y.; Maruyama, K.; Matsumoto, K. *Tetrahedron Lett.* **1984**, *25*, 1075.

110. Aben, R. W. M.; Smith, R.; Scheeren, H. W. *J. Org. Chem.* **1987**, *52*, 365.

111. Chmielewski, M.; Kaluza, Z.; Belzecki, C.; Salanski, P.; Jurczak, J. *Tetrahedron Lett.* **1985**, 2441; *Ibid.* **1984**, 4797.

112. Dauben, W. G.; Krabbenhoft, H. O. *J. Am. Chem. Soc.* **1976**, *98*, 1992.

113. See, for example, Matsumoto, K. *Synthesis* **1985**, 999 and references therein.

114. Jurczak, J.; Kozluk, T.; Filipek, S.; Eugster, C. H. *Helv. Chim. Acta* **1983**, *66*, 222.

115. Jurczak, J.; Kawczynski, A. L., Kozluk, T. *J. Org. Chem.* **1985**, *50*, 1106.

116. Tian, G. R.; Sugiyama, S; Mori, A.; Takeshita, H. *Chem. Lett.* **1987**, *8*, 1557.

117. Barlow, M. G.; Haszeldine, R. N.; Hubbard, R. J. *J. Chem. Soc. Chem. Commun.* **1969**, 301.

118. Lee, C. K.; Hahn, C. S.; Noland, W. E. *J. Org. Chem.* **1978**, *43*, 3727.

119. Kotsuki, H.; Mori, Y.; Nishizawa, H.; Ochi, M.; Matsuoka, K. *Heterocycles* **1982**, *19*, 1915.

120. Lee, C. K.; Hahn, C. S.; Noland, W. E. *J. Org. Chem.* **1978**, *43*, 3727.

121. Drew, M. G. B.; George, A. V.; Isaacs, N. S.; Rzepa, H. S. *J. Chem. Soc. Perkin Trans. I* **1985**, 1277.

122. This observation contrasts with endo/exo studies of Diels–Alder reactions at 1 kbar, which correlate an increase in endo/exo ratio with an increase in solvent polarity. See Berson, J. A.; Hamlet, Z.; Mueller, W. A. *J. Am. Chem. Soc.* **1962**, *84*, 297.

123. Jenner, G. *Tetrahedron Lett.* **1987**, *28*, 3927.

124. (a) Jurczak, J.; Chmielewski, M.; Filipek, S. *Synthesis* **1979**, 41. (b) Raifel'd, Yu. E.; El'yanov, B. S.; Makin, S. M. *Izv. Akad. Nauk SSSR, Ser. Khim.* **1976**, 1090; *Engl. Trans.* **1976**, 1058 (*C.A.* **1976**, *85*, 123,233).

125. Golebiowski, A.; Izdebski, J.; Jacobson, U.; Jurczak, J. *Heterocycles* **1986**, *24*, 1205. See also, Jurczak, J. *Physica* **1986**, *139* and *140B*, 709 and references therein.

126. Umezana, S. *J. Org. Synth. Chem. Jpn.* **1982**, *40*, 1051.

127. For use of Eu(fod)$_3$ as catalyst in similar reactions at atmospheric pressure, see Castellino, S.; Sims, J. J. *Tetrahedron Lett.* **1984**, *25*, 2307.

128. (a) Jurczak, J.; Bauer, T.; Kihlberg, J. *J. Carbohydr. Chem.* **1985**, *4*(3), 447. (b) For a review of the use of 2,3-O-isopropylideneglyceraldehyde in synthesis, see Jurczak, J.; Pikul, S.; Bauer, T. *Tetrahedron* **1986**, *42*, 447.

129. For a thorough study on the effects of high pressure on the Diels–Alder reaction of 1-methoxybuta-1,3-diene and 2,3-O-isopropylidene-D-glyceraldehyde, see Jurczak, J.; Bauer, T. *Tetrahedron* **1986**, *42*, 5045.

130. (a) Jurczak, J.; Bauer, T.; Jarosz, S. *Tetrahedron* **1986**, *42*, 6477. (b) Jurczak, J.; Bauer, T.; Jarosz, S. *Tetrahedron Lett.* **1984**, *25*. 4809. (c) For related work see Jurczak, J.; Bauer, T. *J. Carbohydr. Res.* **1987**, *160*, C1.

131. Tietze, L. F.; Hubsch, T.; Voss, E.; Buback, M.; Tost, W. *J. Am. Chem. Soc.* **1988**, *110*, 4065.

132. Tietze, L. F.; Voss, E. *Tetrahedron Lett.* **1986**, *27*, 6181.

133. Dauben, W. G.; Kessel, C. R.; Takemura, K. H. *J. Am. Chem. Soc.* **1980**, *102*, 6893.

134. Li, T.; Wu, Y. L. *J. Am. Chem. Soc.* **1981**, *103*, 7007.

135. Chmielewski, M.; Jurczak, J. *J. Org. Chem.* **1981**, *46*, 2230.

136. Smith, A. B., III; Liverton, N. J.; Hrib, M. J.; Silvaramakrishnan, H.; Winzenberg, K. *J. Am. Chem. Soc.* **1986**, *108*, 3040.

137. Daniewski, W. M.; Kubak, E.; Jurczak, J. *J. Org. Chem.* **1985**, *50*, 3963.

138. Dauben, W. G.; Gerdes, J. M.; Smith, D. B. *J. Org. Chem.* **1985**, *50*, 2576.

139. (a) Ruzicka, L.; Lardon, F. *Helv. Chim. Acta* **1946**, *29*, 912. (b) Lederer, E.; Marx, F.; Mercier, D.; Perot, G. *Ibid.* **1946**, *29*, 1354.

140. (a) Giguere, R. J.; Bray, T. L.; Duncan, S. M.; Majetich, G. *Tetrahedron Lett.* **1986**, *27*, 4945. (b) Giguere, R. J.; Bray, T. L., Duncan, S. M.; Majetich, G. *The Second Symposium on the Latest Trends in Organic Synthesis*, Virginia Tech, Blacksburg, VA, May 1986.

141. (a) Gedye, R.; Smith, F.; Westaway, K.; Humera, A.; Baldisera, L.; Laberge, L.; Rousell, J. *Tetrahedron Lett.* **1986**, *27*, 279. (b) Gedye, R. N.; Smith, F. E.; Westaway, K. C. *Can. J. Chem* **1987**, *66*, 17.

142. Hasek, J. A.; Wilson, R. C. *Anal. Chem.* **1975**, *46*, 1160.

143. (a) Kingston, H. M.; Jassie, L. B. *Anal. Chem.* **1986**, *58*, 2534. Other chemical applications of microwave heating include (b) rapid determination of thermodynamic functions of chemical reactions: Bacci, M.; Bini, M.; Checcucci, A.; Ignesti, A.; Millanta, L.; Rubino, N.; Vanni, R. *J. Chem. Soc. Faraday Trans. 1* **1981**, *77*, 1503; (c) microwave-induced polymerization: Vanderhoff, J. W., U.S. Patent 3,432,413 [*C.A.* **1969**, *70*, 97422v]; (d) wet ashing procedures: Keyzer H. *Chem. Aust.* **1978**, *45*, 44. Nadkarni, R. A. *Anal. Chem.* **1984**, *56*, 2233. Barrett, P.; Davidowski, L. J.; Penaro, K. W.; Copeland, T. R. *Anal. Chem.* **1978**, *7*, 1021 and references therein; (e) catalytic hydrogenation of alkenes: Van, J. K. S.; Wolf, K.; Heyding, R. D. In *Catalysis on the Energy Surface*, Kaliaguine, S., Mahay, K., eds., Elsevier Science Publishers, Amsterdam, 1984, pp. 561–568.

144. Walkins, K. *J. Chem. Ed.* **1983**, *60* (12), 1043.

145. For a discussion of microwave heating, see Copson, D. A., *Microwave Heating*, The AVI Publishing Company, Westport, CT, 1975, 2nd ed., pp. 8–18.

146. The equation for molar polarizability is

$$P_{\mathrm{m}} = \frac{D-1}{D+2}\ \frac{M}{\rho} = \frac{4\pi}{3}\,N\ \left(\alpha + \frac{\mu^2}{3kT}\right)$$

where D is dielectric constant, M is molecular weight, ρ is density, α is polarizability, μ is dipole moment, and T is temperature.

147. A double-walled ampule ($\sim 10\,\mathrm{ml}$ volume) is charged with $2.0\,\mathrm{ml}$ of solvent, degassed, and sealed.

148. The author is indebted to Mr. George Berman (Custom Cabinet, Macon) for suggestion of Corian as well as design and construction of these containers.

149. *Organikum*, VEB Deutscher Verlag, Berlin 1977, p. 352.

150. Backmann, W. E.; Scott, L. B. *J. Am. Chem. Soc.* **1948**, *70*, 1458.

151. Lohaus, H. *Liebigs Ann. Chem.* **1935**, *516*, 295.

152. McCulloch, A. W.; Stanovnik, B.; Smith, D. G.; McInnes, A. G. *Can. J. Chem.* **1969**, *47*, 4319.

153. Bartlett, P. D.; Woods, G. F. *J. Am. Chem. Soc.* **1940**, *62*, 2933.

154. These products were isolated by chromatography; distillation undoubtedly causes retro-Diels–Alder reaction, hence the low literature yields. The product of entry 5 presumably also undergoes retro-Diels–Alder reaction under microwave conditions.

155. Claisen, L.; Eisleb, O.; Kremers, F. *Liebigs Ann. Chem.* **1919**, *418*, 69.

156. Hoffmann, W.; Pasedach, H.; Pommer, H.; Reif, W. *Liebigs Ann. Chem.* **1971**, *747*, 60.

157. (a) Giguere, R. J.; Lopez, B.; Namen, A.; Majetich, G. Presented at the 38th South-eastern Regional Meeting of the ACS, Louisville, KY November, 1986. Abstract No. ORGN 216. (b) Giguere, R. J. "Microwave Heating in Organic Synthesis," an invited symposium in *Recent Developments in Organic Synthesis*, presented at the 39th South-eastern Regional Meeting of the ACS, Orlando, FL, November 1987.

158. (a) Dissertation (Dr. rer. nat.) of C.-H Bong, Universität Köln, West Germany, 1952. The author thanks Dr. Michael Boberg (Bayer AG, Wuppertal) for kindly obtaining a copy of Bong's thesis. (b) In 1966, this reaction was reinvestigated but no yield was reported: see Huebner, C. F.; Donoghue, E.; Dorfman, L.; Stuber, F. A.; Danieli, N.; Wenkert, E. *Tetrahedron Lett.* **1966**, 1185.

159. Giguere, R. J.; Namen, A. M.; Lopez, B.; Arepally, A.; Ramos, D. E.; Majetich, G.; Defauw, J. *Tetrahedron Lett.* **1987**, *28*, 6553. For studies on tandem ene/intramolecular Diels–Alder reactions with phosphaalkyne enophiles, see Fuchs, E. P. P.; Rosch, W.; Regitz, M. *Angew. Chem. Int. Ed. Engl.* **1987**, *26*, 1011.

160. Alder, K.; von Brachel, H. *Liebigs Ann. Chem.* **1962**, *651*, 141.

161. Reaction of ergosteryl acetate and acrylonitrile at 130°C affords ene adducts in which the carbon–carbon bond is formed at the α-carbon of the enophile, consistent with our observation; see Jones, D. N.; Greenhalgh, P. F.; Thomas, I. *Tetrahedron* **1968**, *24*, 5215.

162. Giguere, R. J.; Majetich, G. Unpublished results 1987.

163. We thank Prof. James Marshall and Barry Shearer of the University of South Carolina for a generous sample of trienal **65** and for subsequent capillary GLC analysis of compounds **66**. Confer reference 164 for the synthesis of **65** and its thermal and Lewis acid-catalyzed Diels–Alder reaction.

164. Marshall, J. A.; Shearer, B. G.; Crooks, S. L. *J. Org. Chem.* **1987**, *52*, 1236.

165. Cf. A. Roedig in Houben-Weyl, 4th ed., Vol. 5/4, 1960, pp. 595–605.

166. Giguere, R. J.; Lopez, B. Unpublished results 1987.

167. Reference 140 prompted over 75 formal reprint requests from 15 countries within 6 months of publication. For a popular article describing microwave heating in synthesis, see *Chem. Br.* **1986**, *22*, 894.

ALLYLSILANES IN ORGANIC SYNTHESIS

George Majetich

OUTLINE

Organic Synthesis: Theory and Application, Vol. 1, pages 173–240.
Copyright © 1989 by JAI Press Inc.
All rights of reproduction in any form reserved.
ISBN: 0-89232-865-7

Anyone who has ever entered Professor Gilbert Stork's office eventually ends up discussing chemistry at his blackboard. It was there that he suggested we complete a synthesis of maytansine via an intramolecular addition of an allylsilane to an aldehyde (Scheme 1). After I failed to respond enthusiastically, I was asked to voice my concerns. Professor Stork begrudgingly acknowledged that it might prove difficult to form a 19-membered lactam using a Lewis acid in the presence of such varied and sensitive functionality. I left his office wondering what chemistry was possible with allylsilanes.[1]

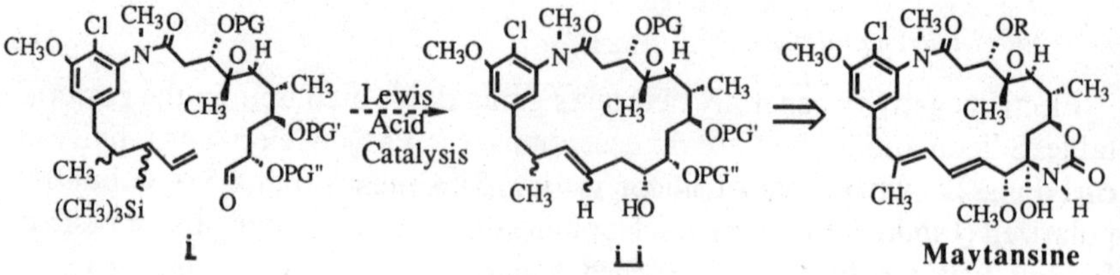

Scheme 1

1. INTRODUCTION

The use of organosilicon reagents in organic synthesis continues to grow at a remarkable rate. This explosion of activity is reflected in the number of books[2] and reviews[3] dedicated to this field of chemistry. The research of Hideki Sakurai, Akira Hosomi, and Ian Fleming has stimulated the use of allylsilanes both as reagents and as intermediates in organic synthesis. These pioneers and others have documented the versatility of this functional group. In contrast to allyl derivatives of most metals, allylsilanes are configurationally stable and relatively inert to oxygen and moisture. These properties allow an allylsilane moiety to be carried through several steps in a synthetic sequence, thereby enhancing its utility.

A. Scope of the Review

Several comprehensive reviews have already appeared, although they deal primarily with allysilane chemistry reported prior to 1981.[4] The myriad of applications and preparations of allylsilanes that have been published since then exceeds the scope of this review. This work focuses only on recent developments in allylsilane chemistry that hold promise for use in organic synthesis.[5] It has been divided into sections on inter- and intramolecular reactions, and each part is further subdivided on the basis of whether addition of the allylsilane generates a carbanionic or carbonium ion intermediate. It is not presented chronologically. Finally, cycloaddition[2b,6] and thermal reactions[2a] of allylsilanes are omitted in this survey.

B. Fundamental Properties of Organosilanes

A brief summary of the physical properties of organosilicon compounds is useful. The reactions of organosilicon compounds are based on silicon's $3s^2 3p^2 3d^0$ valence configuration, analogous to carbon's $2s^2 2p^2$. The important differences between carbon and silicon are the size of the bonding orbitals, the electronegatives, and the availability of silicon's vacant $3d$ orbitals for bonding. These factors enable silicon to undergo nucleophilic substitution and to stabilize cations in a β position to it and anions α to it.

1. Nucleophilic Substitution

Electronegativity generally decreases going down a column in the periodic table. Silicon and carbon are no exceptions; the Pauling electronegativity of carbon is 2.5, where that of silicon is 1.8.[7] This means that a Si—C bond is polarized (Figure 1), making nucleophilic attack at the silicon atom possible. It is in fact favored if the expelled fragment is a good leaving group.[3c]

Nevertheless, organosilanes are relatively nonpolar compared to other organometallic compounds, so they can tolerate the presence of other functional groups and can be handled without special precautions.

$$\overset{\delta^+}{Si} \longrightarrow \overset{\delta^-}{C}$$

Figure 1

Bonds between silicon and electronegative elements, such as the halogens, are stronger than expected (Table 1).[8] The driving force of many organosilane reactions is the formation of a compound with the stronger bond to silicon. For example, treatment of trimethylsilyl ether (1) with fluoride ion

Table 1.[8]   Bond Engeries (kJ/mol)

C—F 552	Si—F 552.7 $\pm$ 2.1
C—Cl 397 $\pm$ 29	Si—Cl 456 $\pm$ 42
C—Br 280 $\pm$ 21	Si—Br 343 $\pm$ 50
C—I 209 $\pm$ 21	Si—I 293

generates fluorotrimethylsilane (2) because the resulting Si—F bond is stronger than the Si—O bond being broken (Scheme 2). The facile removal of a silyl group by fluoride ion is commonly referred to as desilylation and is responsible for the extensive use of silicon reagents as protecting groups.[9]

$$ROSi(CH_3)_3 \xrightarrow{\ F^-\ } F\text{-}Si(CH_3)_3 \ + \ RO^-$$

1   **2**   **3**

Scheme 2

In the desilylation shown above, the Si—O bond was broken rather than a Si—C bond because an alkoxide (cf. 3) is a better leaving group than a methyl anion. Treatment of carbon-functionalized silanes with hard bases such as fluoride ion or sterically unencumbered oxygen nucleophiles,[10,11] however, generates anionic species that behave as carbanions (Scheme 3). The nature of this nucleophile is addressed later in the section on fluoride ion-promoted reactions of allylsilanes.

$$R_4Si \xrightarrow{\ F^-\ } R_3SiF \ + \ R^-$$

R = alkyl, allyl, or aryl substituents

Scheme 3

Nakamura and Kuwajima[12] were among the first to exploit the use of organosilanes as masked carbanions. In 1976 they produced an acetylide anion via fluoride ion-promoted desilylation of a trimethylsilyl acetylene (**4**),

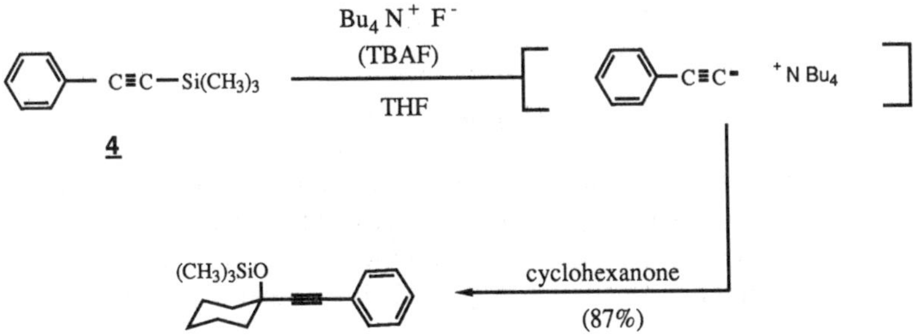

Scheme 4[12]

which reacted with aldehydes and ketones to yield substituted propargyl alcohols in good yields (Scheme 4). Similar results were independently obtained by Holmes and co-workers, who formed the monoacetylide of *bis*-trimethylsilylacetylene (**5**) with KF and 18-crown-6 (Scheme 5).[13] The potential of the C–SiR$_3$ moiety to function as a protected carbanion is reviewed elsewhere.[14]

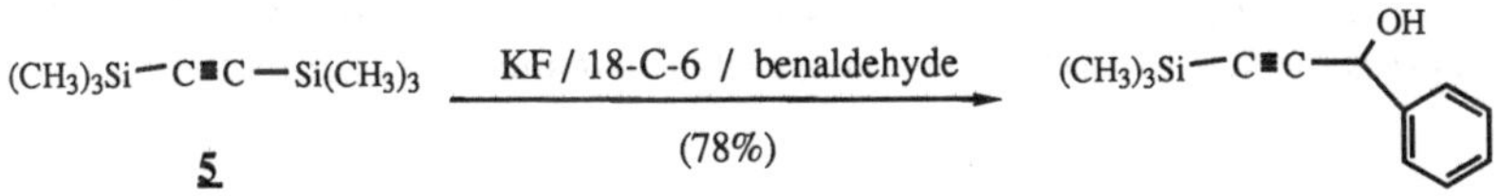

Scheme 5[13]

2. α-Metallated Organosilanes

In contrast to halides or alkoxides, carbon nucleophiles such as alkyllithiums may act as nucleophiles or bases toward alkylsilanes. For example, *n*-butyllithium attacks the silicon atom of chloromethyl trimethylsilane (Scheme 6, **6** → **7**),[15] whereas *sec*-butyllithium abstracts a methylene proton from **6** to furnish an α-metallated organosilane (**8**). Even though silicon is electropositive, it is still able to stabilize an adjacent carbanion, although the reason for this stabilization is still unclear and a subject of discussion.[16] Metallated alkylsilanes are key reagents in silyl–Wittig reactions.[17]

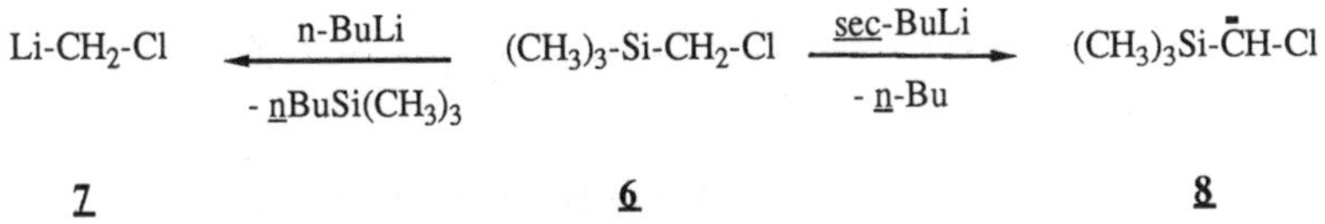

Scheme 6[15]

3. β-Stabilized Carbonium Ions

Reactions that involve the formation of a carbonium ion β to the silicon atom exhibit unusually high reactivity and regioselectivity. This behavior was predicted in a theoretical treatment by Sommer and Whitmore[18] and experimental evidence supports this conclusion. For example, Eaborn and co-workers demonstrated that 1-chloro-2-trimethylsilyl ethane (9) solvolyzes faster than t-butyl chloride (Scheme 7).[19] Clearly the presence of silicon stabilized the primary carbonium ion, although the exact nature of this stabilization is still under debate.[20]

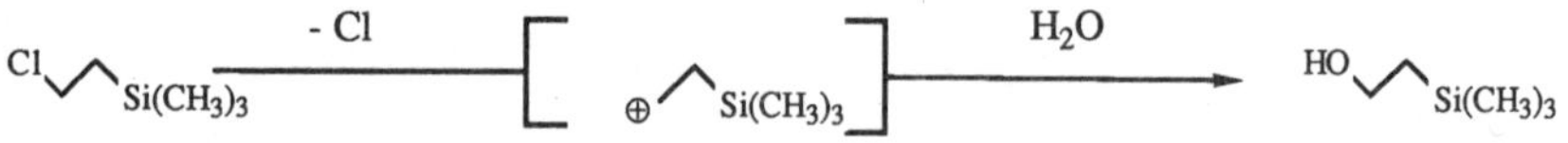

2   iii

Scheme 7

The reader is referred elsewhere[2,3] for more comprehensive accounts of the generalizations mentioned above. The reactions of allylsilanes that follow are primarily based on silicon's ability to undergo nucleophilic substitution and to stabilize a β-carbonium ion.

II. INTERMOLECULAR ALLYLSILANE REACTIONS

A. Electrophilic Substitution Reactions

The stabilization of a β-carbonium ion by silicon is the key for a variety of useful electrophilic substitution reactions. During the past decade the electrophilic substitution of allylsilanes was thoroughly studied from both synthetic and mechanistic standpoints. For convenience, these reactions have been organized into three categories based on the nature of the electrophile.

1. Protic Acids

The susceptibility of allylsilanes to electrophilic attack was first recognized in 1948 by Sommer and Whitmore.[18] They correctly theorized that allyltrimethylsilane (10) would react with an electrophile to give a silicon-stabilized cationic intermediate (iv), which would then undergo nucleophilic attack on silicon to give an allylated adduct, i.e., (**11**). Note that the carbon–carbon double bond migrates in this reaction (Scheme 8).

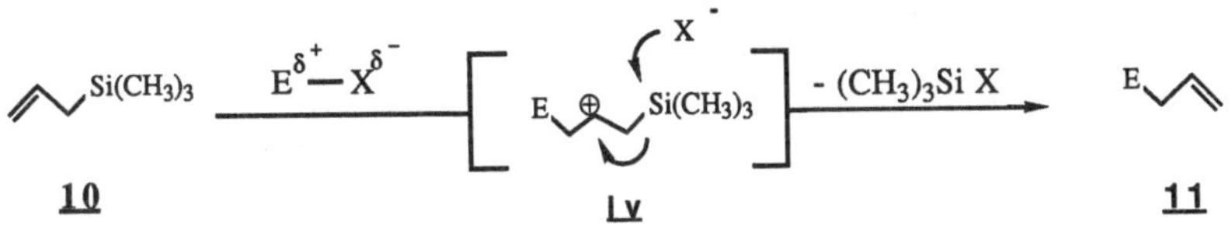

Scheme 8

As an extension of these results, allyltrimethylsilane under protic acid conditions yielded the regioisomeric desilylated material (**12**) with hydrochloric acid and sulfuric acids, whereas hydrobromic and hydroiodic acids underwent Markovnikov addition to the double bond, cf. **13** (Scheme 9).

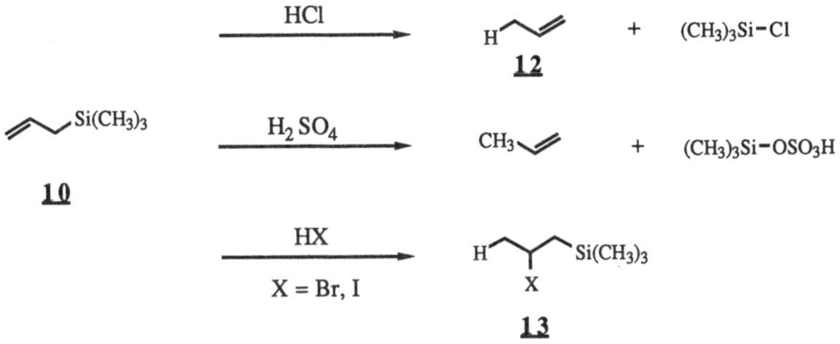

Scheme 9

These results reflect the relative rates of attack by halide or sulfate ions on the carbon and silicon atoms of the cationic intermediate (cf. **iv**, Scheme 8).[22] Attack on silicon by chloride and sulfate ions leads to stronger silicon–oxygen and silicon–chlorine bonds relative to silicon bonds with bromine and iodine. The reaction of a protic acid with an allylsilane to give desilylated material with transposed regiochemistry of the alkene is known as protodesilylation. Moreover, extensive work has shown that a mixture of boron trifluoride and acetic acid is the preferred reagent for such protodesilylations.

Chart 1 contains four representative examples of protodesilylation. Because of the mechanistic requirements for electrophilic substitution, treatment of silanes **14–17** with protic acid furnishes adducts having a double bond in a specific location.[23,24] This predictability has considerable synthetic potential.

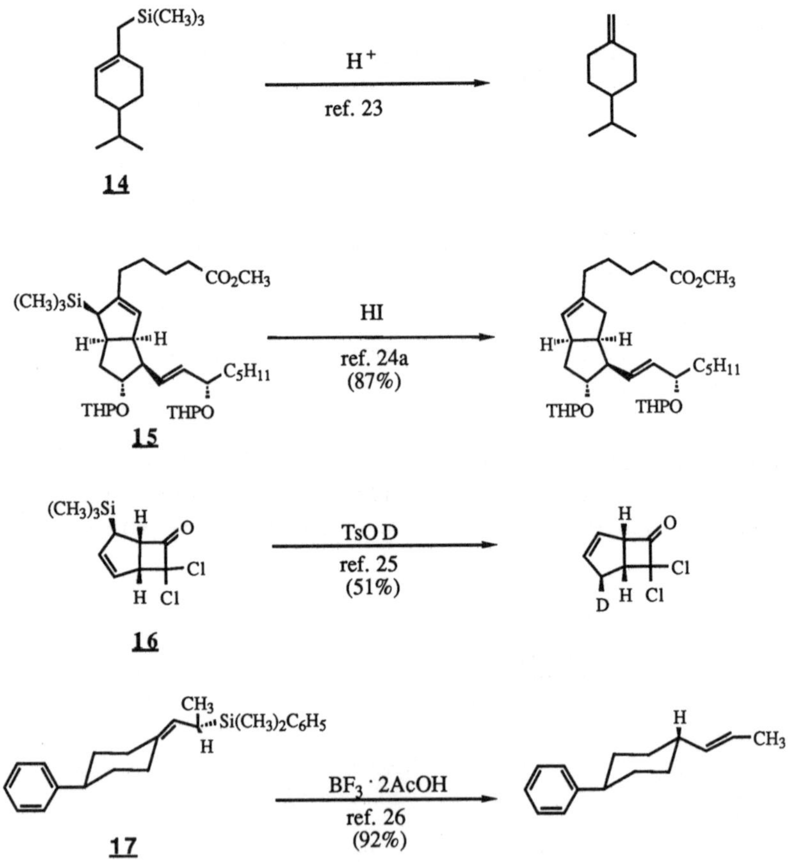

Chart 1

Protodesilylation is also useful as a mechanistic probe of the electrophilic substitution of allylsilanes. For example, Fleming and Au-Yeung addressed the stereochemical requirements of the allylic silicon–carbon bond in the reaction of allylsilane **16** with deutero toluenesulfonic acid (Chart 1).[25] In acyclic systems, allylsilanes react with a wide range of electrophiles with high *anti* selectivity due to the steric bulk of the silyl moiety that dictates the direction of electrophilic attack (cf. the deuterodesilylation of **17**).[26,27] However, Fleming's work with cyclic allylsilanes, such as **16**, revealed that electrophiles can also add *syn* to the silyl moiety (Scheme 10). This substrate dependency was shown to be governed by the requirement that the allylic silicon–carbon bond assumes a stabilizing coplanar relationship with the empty *p* orbital (cf. **v** and **vi**).

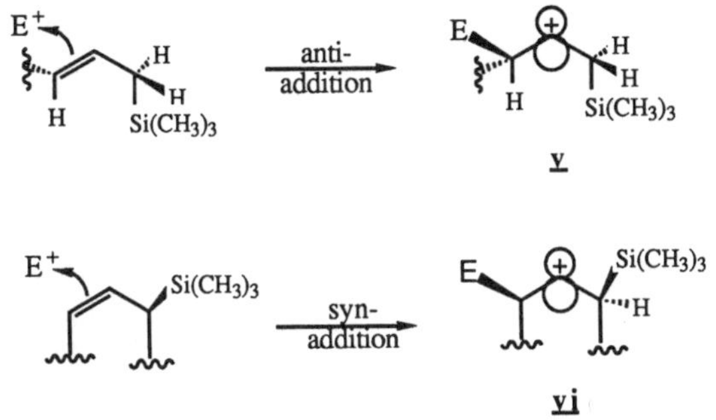

Scheme 10

Hosomi and Sakurai recognized the ability of allylsilanes to react with alcohols to yield trimethylsilyl ethers and propene.[28] This silylation requires the use of a catalytic quantity of either *p*-toluenesulfonic acid or iodine (Scheme 11).

Scheme 11[28]

2. Carbon Electrophiles

In addition to protons, transient cationic species also react with allylsilanes. Indeed a large portion of allylsilane chemistry involves the allylations of reactive species produced when carbonyl compounds, acetals, ketals, or Michael acceptors are activated by Lewis acids.

a. Carbocations. Although allylsilanes react with a wide variety of cations, there are few examples with simple carbonium ions as the reactive electrophilic species. Fleming and others[29] have reported the use of primary, secondary and tertiary alkyl cations to produce allyl-substituted alkanes (Scheme 12). These allylations require reactive alkyl halides, such as *t*-butyl chloride or *t*-amyl chloride, whereas less reactive electrophiles (e.g., iso-propyl chloride) fail to react. The reaction of the tertiary carbonium ion generated from adamantyl chloride (**19**)[29d,e] or from other adamantyl derivatives[29f] represents one such example.

Scheme 12²⁹

Resonance-stabilized cations have also been examined. Hosomi and Sakurai have reported the facile allylic coupling of allylic ethers or halides activated by titanium tetrachloride with allyltrimethylsilane to give good yields of 1,5-dienes (Scheme 13).[30]

Scheme 13³⁰

Mohan and Katzenellenbogen have exploited the ability of aryl groups to stabilize cations in the key step of a synthesis of a hexestrol derivative as shown in Scheme 14.[31] Treatment of benzylic methyl ether **21** with titanium tetrachloride generates a benzylic carbonium ion, which is captured by the substituted allylsilane **22** to furnish alkane **23**.

Scheme 14³¹

Finally, Eguchi and co-workers have coupled allyltrimethylsilane with a cyclopropylmethyl cation to produce two allylated products (in a 1:3 ratio) in which ring opening of the cyclopropane unit had occurred (Scheme 15).[32]

Reaction conditions were developed whereby adduct **26** could be generated regioselectively and in high yield.

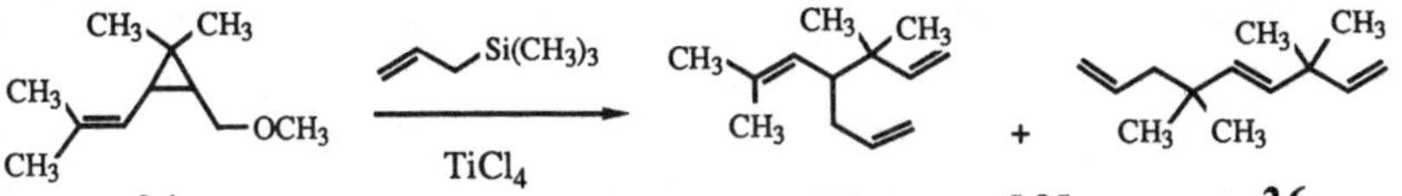

24 25 26

Scheme 15[32]

b. Acetals and Ketals. Dihetero-substituted methyl or methylenes undergo allylation with allylsilanes when a cationic intermediate is formed by acid-catalyzed elimination of one of the hetero substituents. Accordingly, acetals and ketals react with great facility. In 1981, Hosomi and Sakurai reported the efficient allylation of acetals with allyltrimethylsilane with iodotrimethylsilane to initiate the reaction (Scheme 16).[33] The use of stronger (but more typical) Lewis acids gave bisallylated products (cf. **28**).[34a]

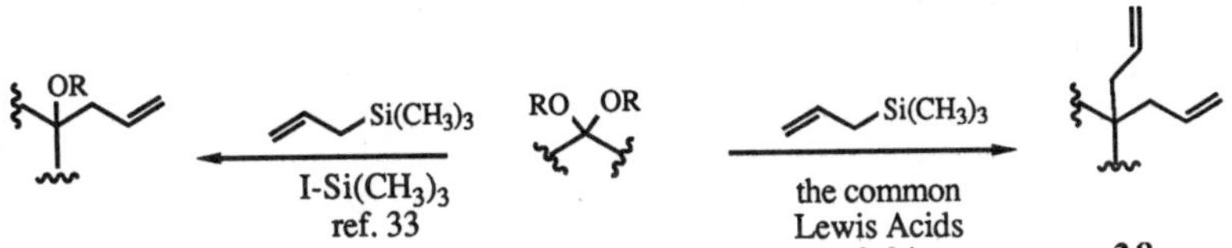

27 28

Scheme 16

Shortly thereafter, the reaction of allylsilanes with acetals was extended to the synthesis of functionalized C-glycoside precursors. As shown in Scheme 17, the C-allylation of glycopyranosyl chlorides and methyl glycopyranosides was achieved stereoselectively with either iodotrimethylsilane or trimethylsilyltrifluoromethanesulfonate to catalyze the reaction.[35]

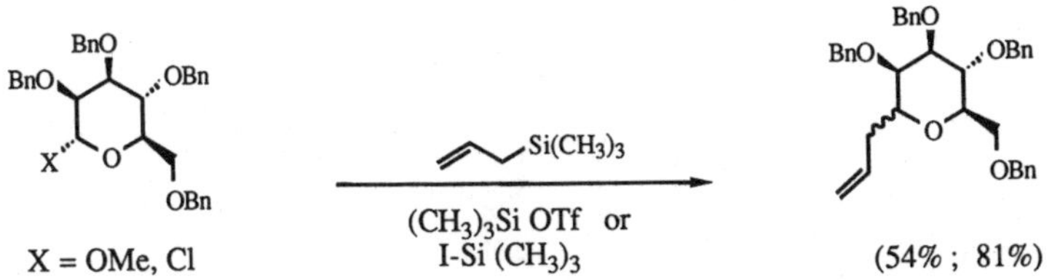

X = OMe, Cl

29

(54% ; 81%)

α / β : 10 / 1

30

Scheme 17[35]

Both Danishefsky and Kerwin[36] and Kozikowski and Sorgi[37] independently investigated the reaction of glycal acetates with allyltrimethylsilane.
These researchers noted that the diastereoselectivity of such allylations is
dependent on both the choice of Lewis acid employed and the nature of the
leaving group of the anomeric center. Kozikowski and Sorgi featured this
allylation reaction in a synthesis of methyl deoxypseudomonate B (35)
whereas Danishefsky and Kerwin applied this process to several glycal
derivatives obtained from natural hexoses (31 → 32, Scheme 18).[38]

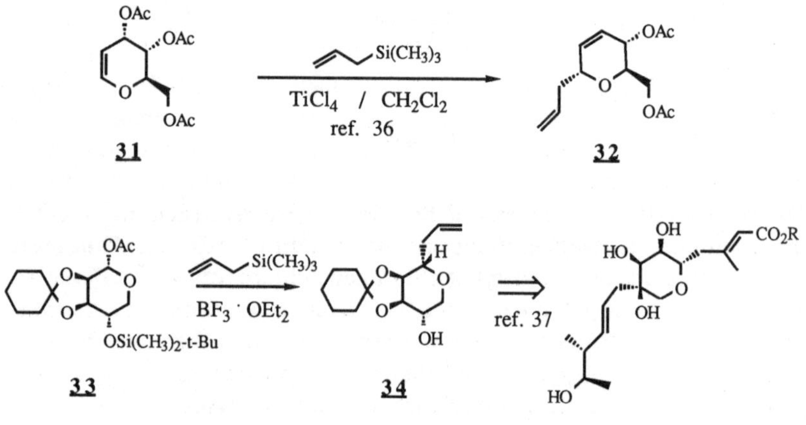

Scheme 18

The use of chiral templates to achieve asymmetric synthesis is now
commonplace. Johnson and Bartlett have studied the Lewis acid-catalyzed
coupling of chiral acetals such as 36 with allyltrimethylsilane to generate
hydroxy ethers in which the new chiral center is formed with high enantio-
selectivity (Scheme 19).[39] Removal of the chiral auxiliary via a two-step
sequence affords homoallylic alcohols (cf. 37) in high optical purity.

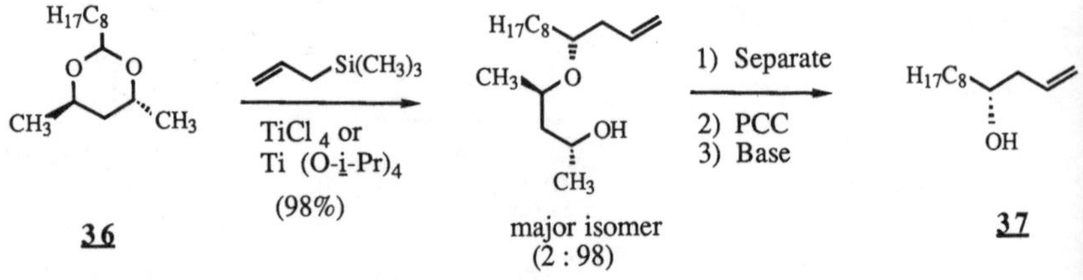

Scheme 19[39a]

Recently Yamamoto and co-workers treated chiral steroidal acetals with allyltrimethylsilane in the presence of titanium tetrachloride and observed that only alcohol **39** was produced, with high asymmetric induction presumably via a concerted mechanism (Scheme 20).[40]

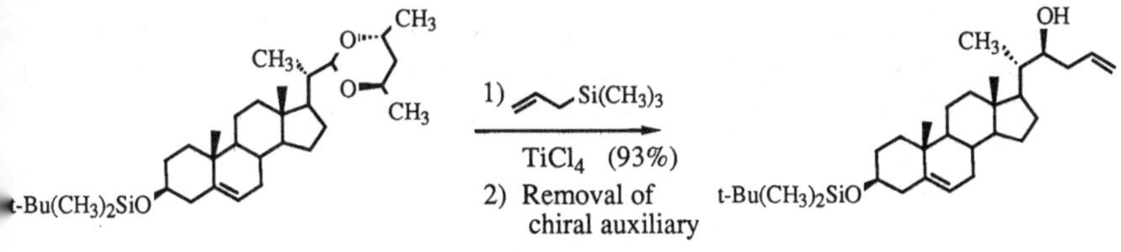

Scheme 20[40]

c. Ketones and Aldehydes. The Hosomi Reaction: The early reactions of allylsilanes dealt primarily with carbonyl compounds. In 1974 Deleris, Dunogues and Calas reported the allylation of chloral and chloroacetone catalyzed by aluminum chloride or gallium chloride.[41] The related reaction of perfluoroacetone (**40**) with allyltrimethylsilane was reported shortly thereafter by Abel and Rowley.[42] Their work showed that perhalogenated ketones are sufficiently polar to react with allylsilanes without catalysis. This reaction, however, varied with temperature (Scheme 21). At higher temperatures, the carbonium ion intermediate is deprotonated intramolecularly to generate vinylsilane **41**, whereas at lower temperatures the cation is quenched by alkoxide to form oxetane **42**. Finally, in the presence of aluminum chloride the alkyl trimethylsilyl ether **43** predominates.

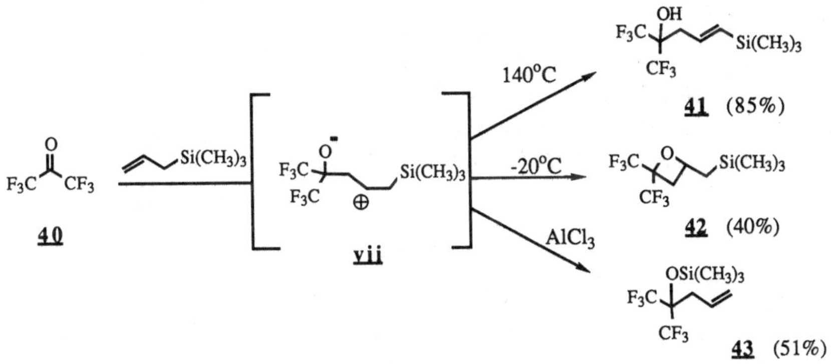

Scheme 21[42]

Shortly after these studies, Hosomi and Sakurai reported that carbonyl compounds of average reactivity react rapidly with allyltrimethylsilane to give γ,δ-unsaturated alcohols (cf. **44**) in excellent yields only if the cationic nature of the carbonyl carbon is enhanced by complexation of the oxygen atom with a Lewis acid such as titanium tetrachloride, aluminum trichloride, stannic chloride, or boron trifluoride etherate (Scheme 22).[43] Although the

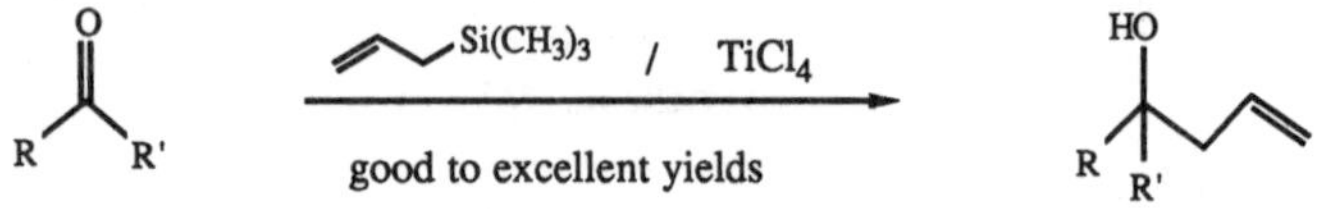

$$
\underset{R \quad R'}{\overset{O}{\|}} \quad \xrightarrow[\text{good to excellent yields}]{\overset{\diagup\!\!\diagup\!\!\diagdown Si(CH_3)_3 \;/\; TiCl_4}{}} \quad \underset{R \quad R'}{\overset{HO}{\diagdown}}
$$

44

Scheme 22[43]

allylation of aldehydes and ketones with allylsilanes was discovered independently by Calas and Sakurai, Hosomi and Sakurai are generally credited with making this allylation procedure a viable synthetic method (see Epilogue). The regioselectivity of this process was demonstrated by the allylation of a carbonyl compound with the isomeric allylsilanes **45** and **47**. The formation of the new carbon–carbon bond in both additions occurred exclusively at the γ-carbon of each allylsilane unit (Scheme 23). As mentioned before, this characteristic reactivity of allylsilanes under electrophilic conditions is dictated by mechanistic considerations.

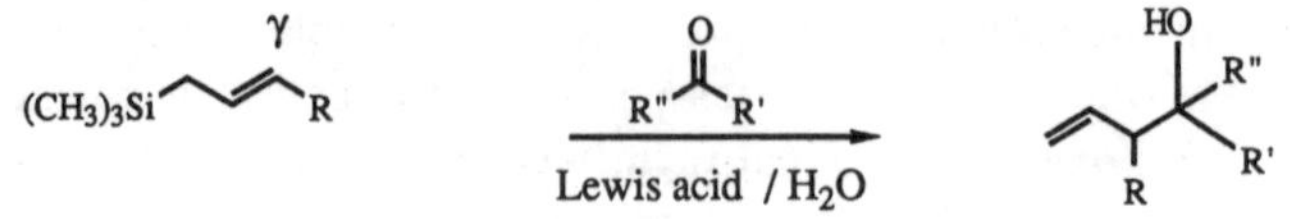

$$(CH_3)_3Si\overset{\gamma}{\diagup\!\!\diagdown}R \qquad \underset{R''}{\overset{O}{\|}}R' \xrightarrow{\text{Lewis acid} \,/\, H_2O} \qquad \text{HO}\, R''$$

45   **46**

$$(CH_3)_3Si\overset{\gamma}{\diagup\!\!\diagdown}R \qquad \underset{R''}{\overset{O}{\|}}R' \xrightarrow{\text{Lewis Acid} \,/\, H_2O} \qquad \text{HO}\, R''$$

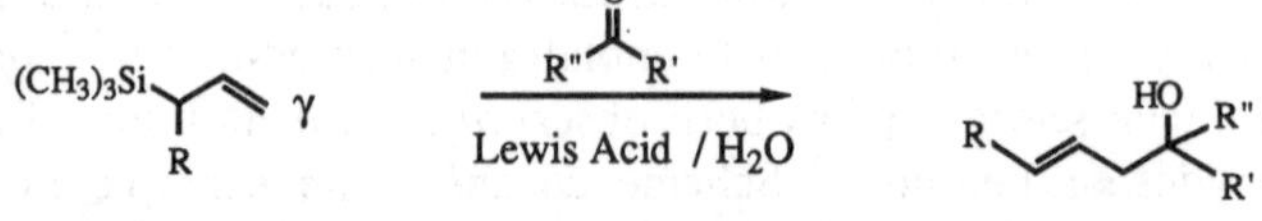

47   **48**

Scheme 23[43]

Soai and Ishizaki have investigated the regioselectivity of allylations of compounds having more than one carbonyl functionality (Scheme 24). They have reported that under Lewis acid catalysis chiral ketone **49** reacts with allyltrimethylsilane exclusively at the ketone carbonyl. Addition of the allylsilane to the less hindered face of the keto-amide chelate accounts for the stereoselectivity of this reaction.[44]

Scheme 24[44]

The regioselectivity of α-keto- and β-ketoacetals has also been studied. Ojima and Kumagai noted that β-ketoacetals, such as **50**, react only at the acetal carbon, independent of the choice of Lewis acid employed.[45] In contrast, α-ketoacetals (i.e., **51**) undergo addition to the carbonyl moiety, although use of strong Lewis acids leads to reaction at both centers (Scheme 25).

Scheme 25[45]

Reliable methods have long been sought to diastereoselectively add allylic nucleophiles to chiral and achiral carbonyl compounds.[46] Although remarkable progress has been made with enolates in such additions,[47] only modest diastereofacial selectivity has been observed for the addition of allylsilanes to aldehydes or ketones.[48] Scheme 26 shows an example in which the diastereofacial preference observed in this allylation is consistent with attack

Scheme 26[48]

of the allylsilane from the less hindered face of the chelated intermediate as predicted by either Cram's or Felkin's model for asymmetric induction.[49] In contrast, the Lewis acid-promoted allylations of α-alkoxy aldehydes (e.g., 53 and 55) and β-alkoxy aldehydes (e.g., 54) proceed with good to excellent diastereofacial selectivity due to chelation control (Scheme 27).[50] Moreover, the degree of 1,2- or 1,3-asymmetric induction observed in these additions was strongly dependent upon the choice of Lewis acid employed.

Scheme 27[50]

The Lewis acid-catalyzed reaction of functionalized allylsilanes with carbonyl compounds has also been studied. Recently Fleming and co-workers reported that optically active heptadienyl silane 57 reacts with isobutyraldehyde and titanium tetrachloride to stereoselectivity give 58 in good yield.[51a] The addition of [β-(alkoxycarbonyl)allyl]silanes to aldehydes enables an efficient, direct preparation of simple α-methylene-γ-butyrolactones (59 → 60).[51b]

(CH₃)₃Si
CO₂CH₃

59

nBuCHO
————————→
TiCl₄ (25 %)
ref. 51b

60

Scheme 28⁵¹

d. Acid Chlorides, Anhydrides, and Esters. Allylsilanes can react with acid chlorides and anhydrides. Although these carbonyl compounds are naturally dipolar, a Lewis acid is nevertheless needed to enhance the electrophilicity of the carbonyl moiety. A classic example of the reaction of an allylsilane with an acid chloride is found in the elegant synthesis of artemesia ketone (**63**) by Dunogues and co-workers (Scheme 29).⁵² It is noteworthy that the reaction of allylsilane **61** with acid chloride **62** occurs via an S$_E$2′ process⁵³ and results in the complete regiochemical transposition of the allyl moiety in excellent yield.

(CH₃)₃Si

61

+

Cl

62

AlCl₃
————————→
(90%)

63

Scheme 29⁵²

The mildness of the reaction conditions is dramatically demonstrated in the aluminum trichloride-catalyzed reaction of α-pinenyltrimethylsilane **64** with acetyl chloride (Scheme 30).⁵⁴ Note that the exocyclic double bond does not isomerize, despite the presence of a strong Lewis acid in the reaction mixture.

Si(CH₃)₃

64

AlCl₃ /
acetyl chloride
————————→
(50%)

COCH₃

65

Scheme 30⁵⁴

As an extension of their work on allylsilanes, Hosomi and Sakurai studied the reactions of pentadienylsilanes, such as **66**, with acid chlorides. They observed that reaction takes place predominantly at the ε-carbon (Scheme 31).⁵⁵

Scheme 31[55]

Meerwein's reagent has been used to induce lactones to react with an allylsilane. For example, lactones **69** and **71**, derived from allylic alcohols, react with allyltrimethylsilane and trimethyloxonium tetrafluoroborate to afford ring-opened products **70** and **72**, respectively (Scheme 32).[56]

Scheme 32[56]

The reaction of allylsilane **73** with the oxonium ion intermediate (cf. **viii**) derived from anhydride **74** is shown in Scheme 33.[57]

Scheme 33[57]

Hosomi and Sakurai have shown that (1-siloxyallyl)silanes (**76**) react with acid chlorides to give γ- ketoaldehydes in good yield (Scheme 34).[58a] Moreover, addition of a new electrophile to the incipient silyl enol ether results in further alkylation.[58b]

Scheme 34[58]

e. α,β-*Unsaturated Enones and Enals. The Sakurai Reaction:* In the reaction with decalone **79**, allylation could occur either in a 1,2- or 1,4-fashion (Scheme 35). In a benchmark study, Hosomi and Sakurai addressed this question and found that allylation occurs only at the β-carbon.[59,60] Since then many other investigators have utilized this novel allylation process. Indeed, the widespread use of the Lewis acid-catalyzed allylation of enones has already resulted in "Name Status" for this single transformation, i.e., the Sakurai reaction (see Epilogue).

Scheme 35[59a]

Chart 2 presents six applications that illustrate the versatility of this methodology.[61-66] The use of 2-chloromethylallyltrimethylsilane (**89**) by Knapp and co-workers warrants special attention because the initial 1,4-allylation produces an intermediate that undergoes an intramolecular alkylation upon treatment with base.[65] This two-step sequence corresponds to a [3 + 2]-annulation of a five-membered ring onto the carbon–carbon double bond of an enone.

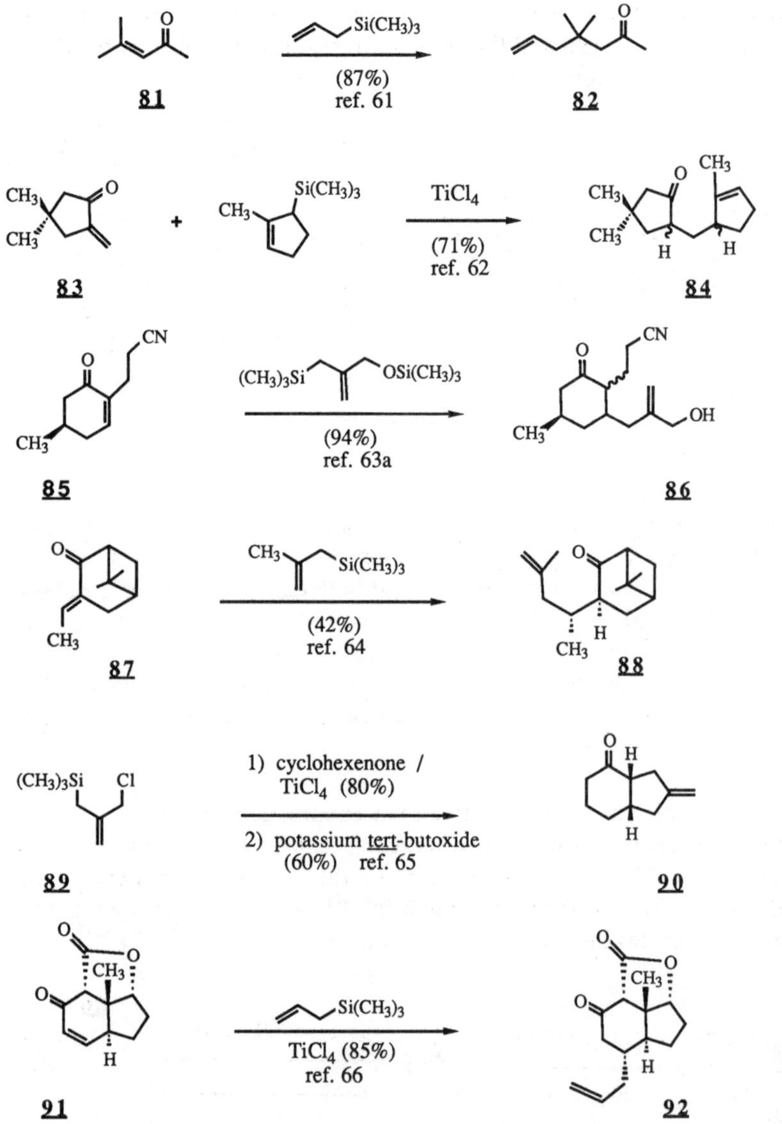

81 → (87%) ref. 61 → **82**

83 + → TiCl₄ (71%) ref. 62 → **84**

85 → (CH₃)₃Si / OSi(CH₃)₃ (94%) ref. 63a → **86**

87 → CH₃ / Si(CH₃)₃ (42%) ref. 64 → **88**

89 (CH₃)₃Si—Cl →
1) cyclohexenone / TiCl₄ (80%)
2) potassium tert-butoxide (60%) ref. 65
→ **90**

91 → Si(CH₃)₃ / TiCl₄ (85%) ref. 66 → **92**

Chart 2

Heathcock and co-workers have reported significant diastereofacial selectivity in their studies of the allylation of acyclic and cyclic enones (Scheme 36).[48,67] For example, Lewis acid-promoted addition of allyltrimethylsilane to enone **93** produces 7:1 mixtures of adducts, with the *anti* isomer (**94**) predominant. The allylation of cyclohexenones **95** and **97** also gives mixtures of 1,4-adducts (cf. **96** and **98**); this diastereoselectivity results from conformational and/or steric control.

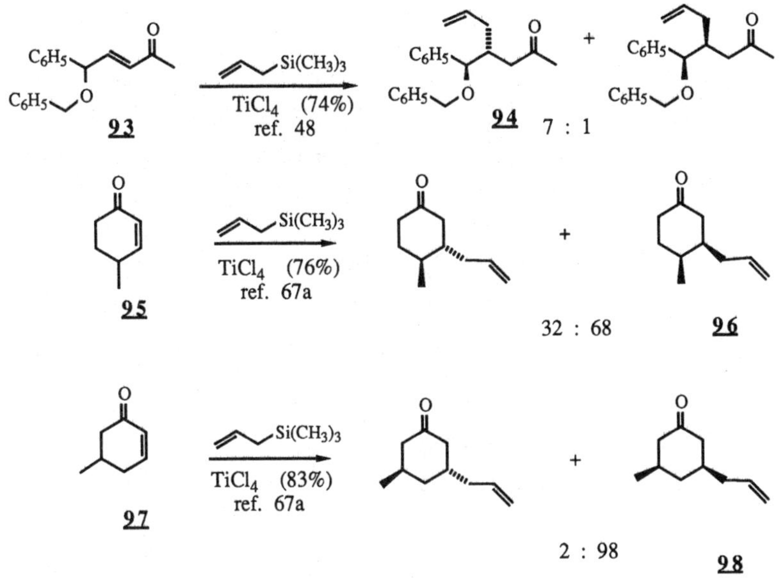

Scheme 36

Enals, which are more reactive than enones, generally react with allyltrimethylsilanes in 1,2-addition when catalyzed by boron trifluoride etherate (Scheme 37).[68] Use of titanium tetrachloride as catalyst, however, results in addition of HCl to form chlorohydrin **101**, which precludes allylation and thus regenerates the starting enal on aqueous work-up.

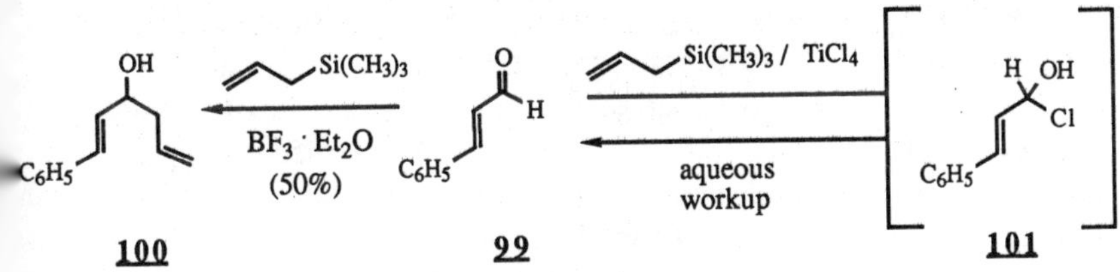

Scheme 37[68]

The initial intermediate in a Sakurai allylation is either a δ,ε-unsaturated titanium trichloride enolate or a δ,ε-unsaturated trimethylsilyl enol ether.[69] Such intermediates are extremely useful, as they can react with acetals or carbonyl compounds in the presence of Lewis acids to yield functionalized ketones (Scheme 38).[70] The use of this simple concept has to date been minimal undoubtedly due to an inability to suppress enolate isomerization (cf. ix → 103).

Scheme 38[69,70]

In 1979 Santelli and co-workers[71] reported that the allylation of enone 104 with allyltrimethylsilane produced both unsaturated ketone 105 and a ketosilane by-product (106) possessing a cyclobutane ring (Scheme 39). This

Scheme 39[71]

observation conflicted with House's earlier report[72] that condensation of 104 with allyltrimethylsilane generated only ketone 105. The contradiction led House and co-workers to reexamine this reaction.[73,74] Control experiments revealed that the production of 106 was governed by the reaction conditions and subsequent work-up. It was determined that significant quantities of silane 106 are produced when enone 104 reacts at elevated temperatures—prior to hydrolysis.[74] House interpreted these results to suggest the following: (1) cationic intermediate x or β-chlorosilane xi is formed during the reaction, prior to work-up (Scheme 40); (2) warming the reaction mixture favors the intramolecular alkylation of the titanium enolate; and (3) addition of water to the reaction mixture at low temperatures forms β-hydroxysilane xii that collapses to unsaturated ketone 105 on warming.

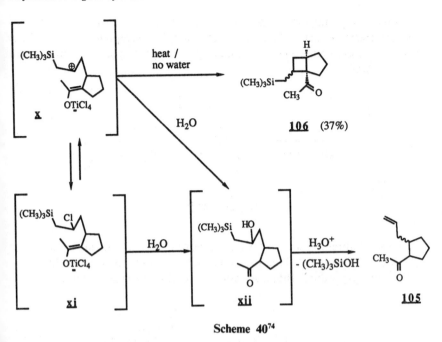

Scheme 40[74]

Santelli and co-workers have studied the allylation of α,β-unsaturated acyl cyanides with *E*-2,4-bis(trimethyl)silyl-2-butene (**107**), a bifunctionalized nucleophile.[75] Their results indicate that the allylsilane moiety that reacts is the one leading to the formation of the more stable carbonium ion (cf. **xiii**). In the example shown in Scheme 41, only product **108** is produced, resulting from the formation of a tertiary, as opposed to secondary, cation. This result clearly indicates that Lewis acid-catalyzed allylations are under thermodynamic control.

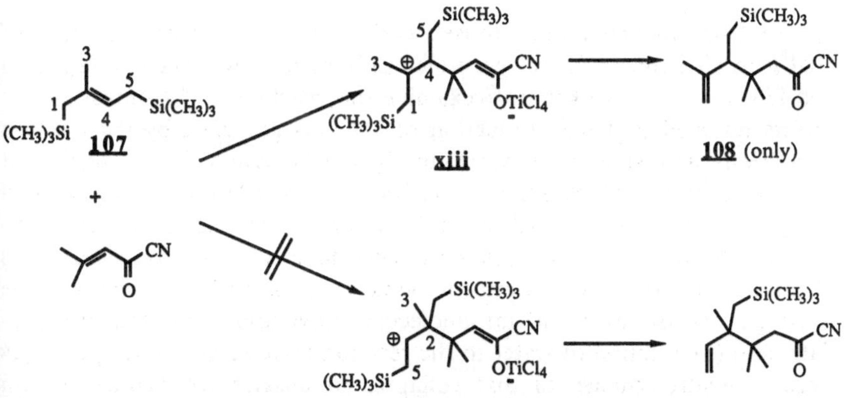

Scheme 41[75]

With quinones as substrates, 1,2-addition of the allyl unit is initially observed.[76] However, the 1,2-adduct undergoes a rapid[1,2]-allyl rearrangement to yield a hydroquinone (cf. 109). 2,6-Dimethoxyquinone 110 was hown to react in 1,2-fashion at the C-1 position (Scheme 42). Vinylogous deactivation of the 4-carbonyl by the methoxy substituents discourages the allyl rearrangement, thus the 1,2-adduct (111) is isolated.

Scheme 42[76]

Direct 1,4-allylation of doubly activated quinones has been observed. In the course of an anthracyclinone study, Murayama regiospecifically allylated acetylbenzoquinone 112 with functionalized allylsilane 113 (Scheme 43).[77] When allyltriphenylsilane was treated with 112, dihydrofuran 115 was produced via intramolecular trapping of the intermediate carbonium ion.

Scheme 43[77]

f. Dienones. The intermolecular 1,6-conjugate addition of allyltrimethylsilane to 3-vinyl-2-cyclohexen-1-one was investigated by Defauw and Majetich.[78] They reported that treatment of 116 with allyltrimethylsilane and titanium tetrachloride at $-78°C$, followed by addition of water to the cold reaction mixture, gave enone silane **117** in 20% yield; enone **118** was also isolated as the major product (Scheme 44). However, warming the reaction mixture to room temperature prior to hydrolysis led to only product **117**, albeit in modest yield. The formation of silane **117** is consistent with a reaction pathway in which the silicon-stabilized cationic intermediate resulting from 1,6-addition of the allylsilane undergoes intramolecular alkylation to form a cyclobutane ring. This analysis complements House's study of the allylation of enone **104** (Scheme 40).[74] More importantly, the observation that allylsilanes react in a 1,6-fashion with conjugated dienones has been used advantageously in an intramolecular context.

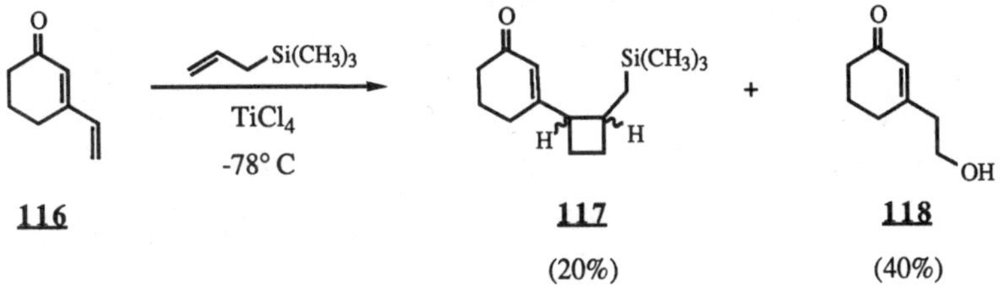

Scheme 44[78]

Recently, Nickisch and Laurent reported comparable observations in the Sakurai allylation of steroidal dienones (**119** → **121**, Scheme 45).[79]

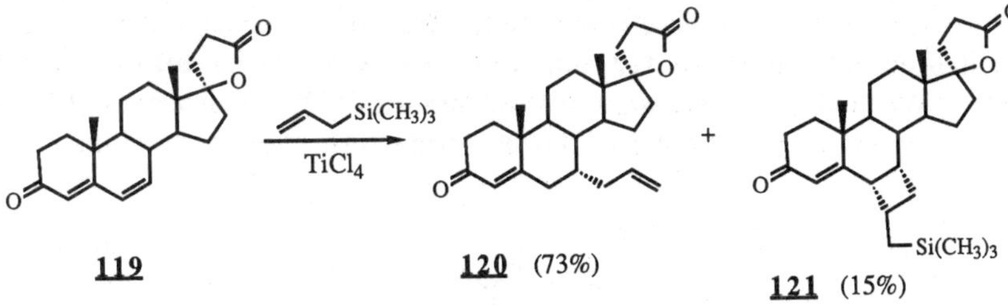

Scheme 45[79]

3. Other Electrophiles

Some species, including ionic and nonionic electrophiles, react with allylsilanes without acidic activation. Westerlund synthesized unsaturated dithianes[80] in good yield by treating allylsilane with 1,3-dithenium tetrafluoroborate (Scheme 46).[81]

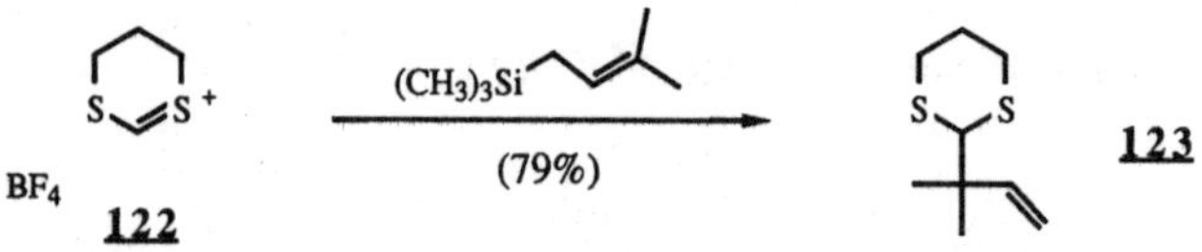

Scheme 46[81]

Though a dithienium salt is discretely ionic, the nonionic chlorosulfonyl isocyanate (124) reacts with allylsilanes to form an O-silyl-N-chlorosulfonyl nitrite (cf. 125) as Fleming showed in his synthesis of loganin 126,[82] or to form a nitrile as reported by Calas and co-workers (61 → 127, Scheme 47).[83]

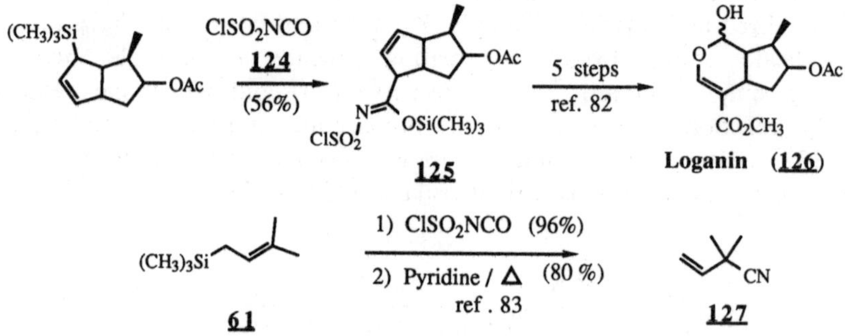

Scheme 47

α-Chlorosulfides undergo facile allylations with allyltrimethylsilane and aluminum trichloride. For example, the allylation of α-chlorosulfide 128 proceeds selectivity, despite the presence of the reactive ketone moiety (Scheme 48).[84] An interesting intramolecular extension of this process is presented later on.

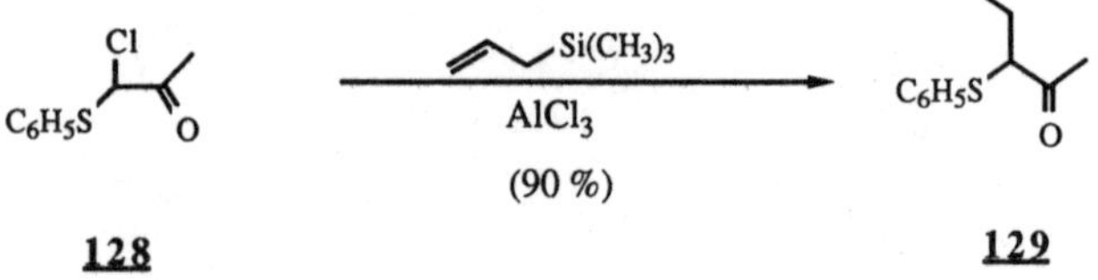

Scheme 48[84]

The wide variety of electrophilic species that react with allylsilanes is further demonstrated by the reaction of iodosobenzene (131), allylsilane 130 and boron trifluoride etherate (Scheme 49). This reaction procedure represents a direct sequence for the regioselective elaboration of allylsilanes to enals.[85]

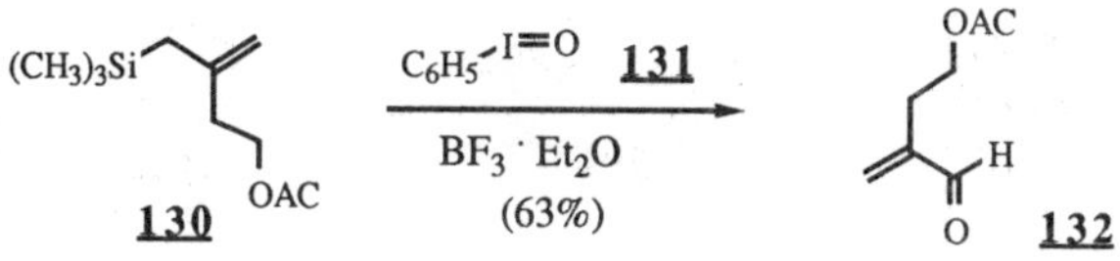

Scheme 49[85]

Oxiranes react with allylsilanes under the influence of Lewis or protic acids to give functionalized pentenols. The conversion of allylsilane **133** to alcohol **134** was first reported by Fleming and Paterson in 1979 (Scheme 50).[86] Since then others have exploited this methodology to prepare functionalized alcohols. For example, Kumada and co-workers have prepared optically active alcohols by treating optically active cyclic allylsilanes with simple oxetanes.[87a] More recently, Carr and Weber have extended this reaction to oxetanes for the preparation of functionalized hexenols (**135 → 136**),[87b] while Oku and co-workers have studied the reaction of 2-vinyltetrahydrofurans with allylsilanes (**137 → 138**).[87c]

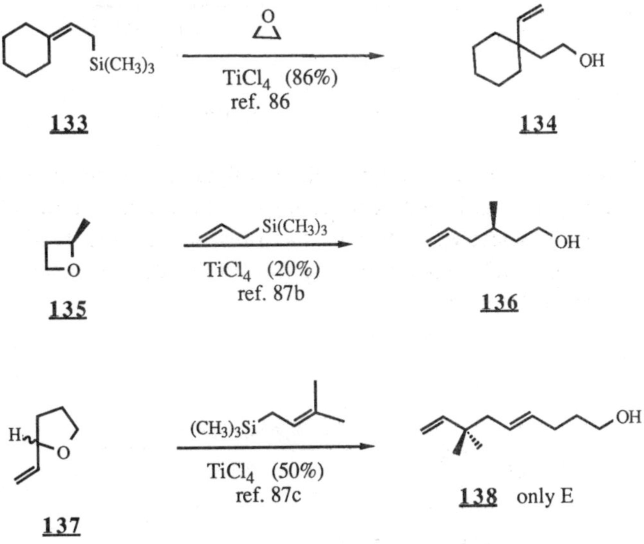

Scheme 50

Nitrogen is an effective stabilizer of α-cations. Accordingly, iminium ions are reactive toward allylsilanes. Thus, elimination of chloride or methoxide from 4-substituted azetidone **139** with catalysis by silver tetrafluoroborate yields an iminium ion that reacts with the allylsilane present in the reaction medium. The stereochemistry of this allylation is controlled by the stereochemistry of the C-3 substituent (Scheme 51).[88]

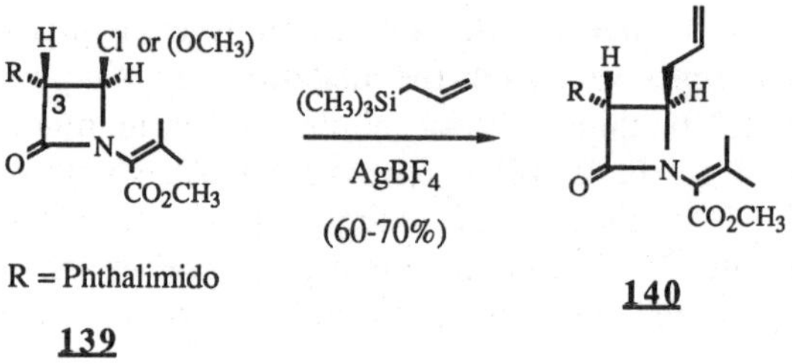

139 → **140**

R = Phthalimido

Scheme 51[88]

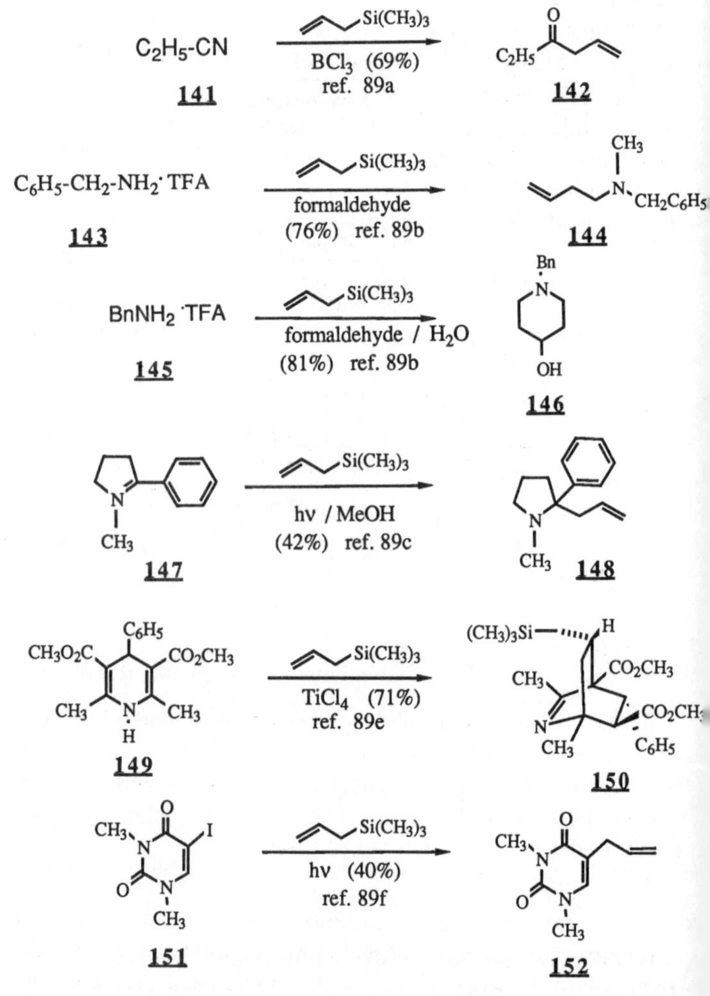

$C_2H_5\text{-}CN$ **141** $\xrightarrow[\text{ref. 89a}]{BCl_3 \ (69\%)}$ **142**

$C_6H_5\text{-}CH_2\text{-}NH_2 \cdot TFA$ **143** $\xrightarrow[\text{(76\%) ref. 89b}]{\text{formaldehyde}}$ **144**

$BnNH_2 \cdot TFA$ **145** $\xrightarrow[\text{(81\%) ref. 89b}]{\text{formaldehyde / } H_2O}$ **146**

147 $\xrightarrow[\text{(42\%) ref. 89c}]{h\nu \text{ / MeOH}}$ **148**

149 $\xrightarrow[\text{ref. 89e}]{TiCl_4 \ (71\%)}$ **150**

151 $\xrightarrow[\text{ref. 89f}]{h\nu \ (40\%)}$ **152**

Chart 3

Chart 3 contains numerous addition reactions with nitrogen based α-cations. These examples vary from the allylation of a simple nitrile to the reaction of simple iminium salts or photolysis of iminium ion salts.[89] Grieco's piperidine synthesis (**145 → 146**) is particularly noteworthy in light of the strongly acidic conditions (TFA) required to achieve formation of the iminium ion in water.[89b] The photocoupling of 6-iodouracil (**151**) to allyltrimethylsilane proceeds via either a radical addition or a carbonium ion mechanism.

The diversity of electrophilic substitution reactions of allylsilanes with nitrogen-based cations is further demonstrated by the addition/elimination or elimination/addition reactions shown in Chart 4.[90]

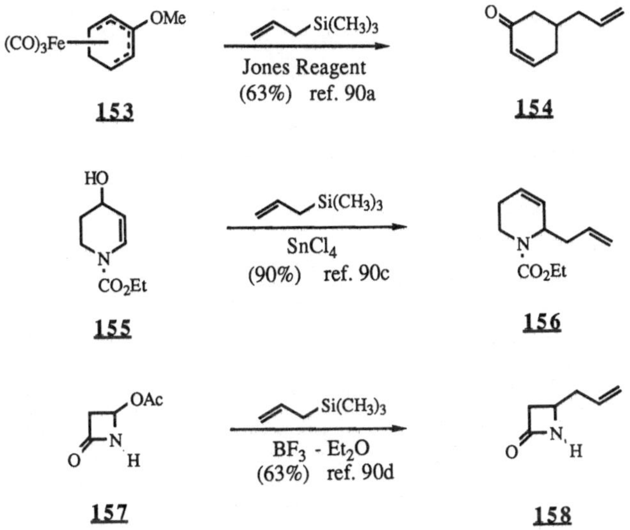

Chart 4

B. Fluoride Ion-Promoted Allylsilane Reactions

Because of the susceptibility of silicon to nucleophilic attack by fluoride ion, treatment of an allylsilane with fluoride ion readily cleaves the carbon–silicon bond to generate an allylic anionic species, which then adds to various carbon electrophiles. The exact nature of this nucleophile is currently under contention.

1. 1,2-Allylations

In 1978 both Sarkar and Andersen[91] and the team of Hosomi, Shirahata, and Sakurai[92] reported a series of fluoride ion-induced allylations of carbonyl compounds (Scheme 52). The nucleophilic species generated failed to react with esters, nitriles, and epoxides. Because Andersen's study dealt with an intramolecular application, it is discussed elsewhere (Section IV.2.c).

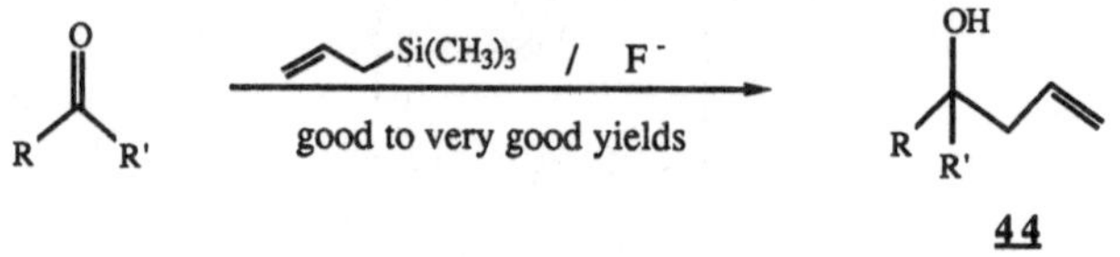

Scheme 52[92]

Three species participate in the fluoride ion-induced allylation procedure: a catalytic quantity of fluoride, allyltrimethylsilane, and a carbonyl substrate. Since a trimolecular reaction is unlikely to occur, a reactive intermediate resulting from the interaction of two of the three components is more plausible (Scheme 53). In their original report of the fluoride ion-induced

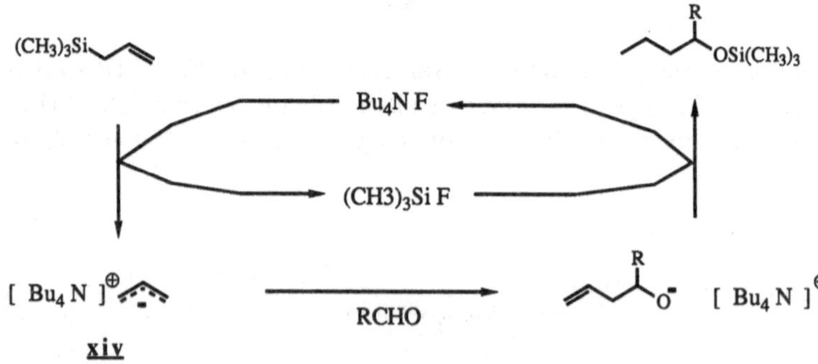

Scheme 53

allylation, Hosomi and Sakurai proposed a catalytic cycle involving a free allyl anion as the reactive intermediate, due to the substantial difference in the Si—C and Si—F bond strengths. Since tetra-n-butylammonium fluoride (TBAF) was used as their fluoride ion source, this neccessitates that the countercation be the tetra-n-butylammonium ion (i.e., **xiv**).[92] Indeed DePuy and co-workers have observed that the gas phase reaction of fluoride ion with allyltrimethylsilane produces an allylic carbanion.[93] In solution, however, the high pK_a of propene, the conjugate acid of an allylic carbanion, makes an allylic anion intermediate quite unlikely. Such qualms led us[68b,94] to speculate that allylfluorosiliconate **xv** is the reactive species (Scheme 54). As mentioned before, participation of the d orbitals permits the valence expansion of silicon; thus hypervalent silicon intermediates such as the

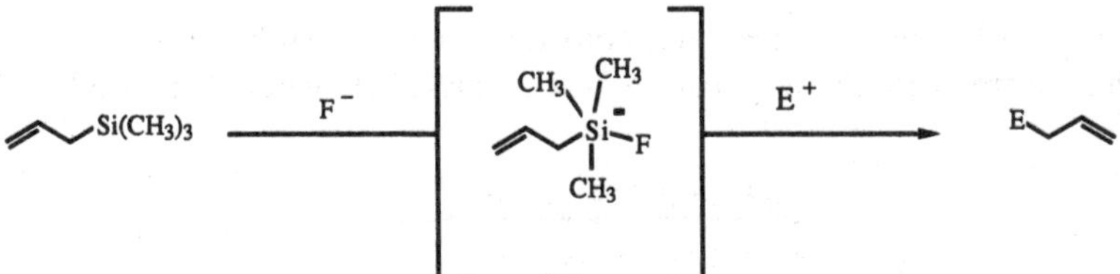

xv

Scheme 54

stable fluorosilicate salt **xvi** (Figure 2) are isolable.[95a] Moreover, penta- and hexacoordinate silicon species are accepted as the reactive intermediates in several organosilicon reactions involving fluoride ion catalysis.[95,96]

SiF_5^-

xvi

Figure 2

Indeed, recently Kira and Sakurai have obtained ^{28}Si-NMR evidence for the intermediacy of a hypervalent allylic silicon intermediate (**xvii**) in the fluoride ion-induced allylation of simple aldehydes (**159** → **160**, Scheme 55).[97a,b]

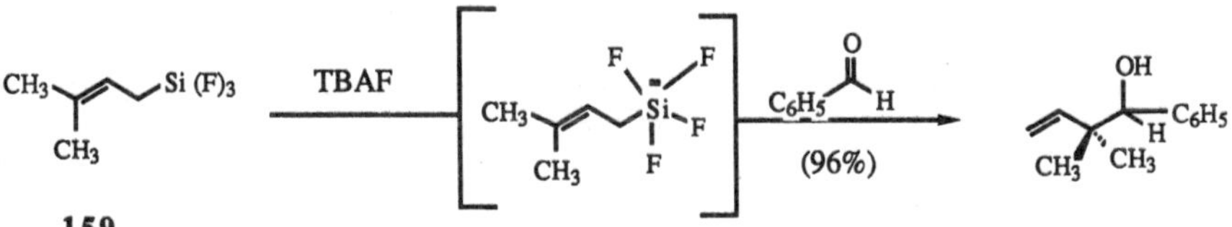

159 **xvii** **160**

Scheme 55[97a,b]

Hosomi has also become an avid proponent of pentacoordinate allylsiliconate nucleophiles. He generated pentacoordinate allylsiliconate **xviii** and reacted it with various simple aldehydes *without catalysis* (Scheme 55).[97c,d] Both Sakurai's and Hosomi's observations strengthen our argument for a

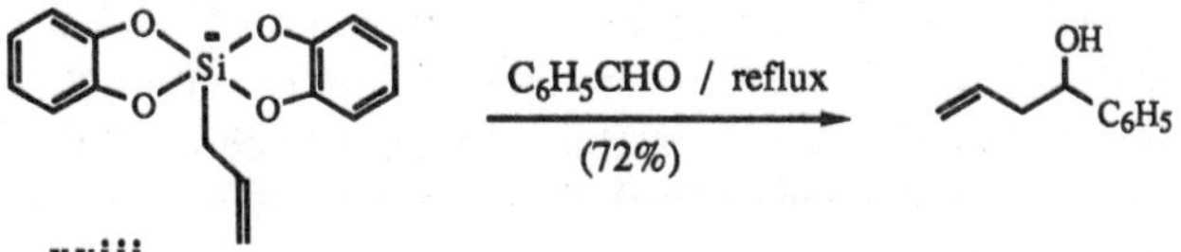

xviii

Scheme 56[97c,d]

tetraalkylfluorosilicate anion in the fluoride-mediated reactions of allyltri-methylsilanes. Nevertheless, it will be extremely difficult to prove that a tetraalkylfluorosilicate nucleophile is involved in these reactions,[98] since replacement of the halogens or the alkoxy substituents (cf. **xvii**) with alkyl groups (cf. **xviii**) will greatly destabilize such anionic species, thus complicating their characterization.

Addition of fluoride ion to an unsymmetrical allylsilane[99] generates an ambident nucleophile that can react at either terminus. The reaction of substituted allylsilanes **161** and **162** with butyraldehyde addressed this question of regiochemistry and resulted in a mixture of γ,δ-unsaturated alcohols **163** and **164** (Scheme 57).[92] Note that a distinct preference is seen for reaction at the primary carbon atom of the allylsilane moiety regardless of the position of the methyl group α or γ to silicon. These results dramatically contrast with the exclusivity of the S_E2' pathway characteristic of the Sakurai allylation.

Scheme 57[92]

The product distribution shown in Scheme 57 indicates that carbon–carbon bond formation occurs at both the α and γ carbons of the π-system. From a mechanistic viewpoint, this process is more appropriately considered as an addition–elimination reaction with the terms S_E2 and S_E2' aptly representing the carbon–carbon bond formation; the addition process refers to the formation of a nucleophilic species via addition of the fluoride ion (Scheme 58).[100]

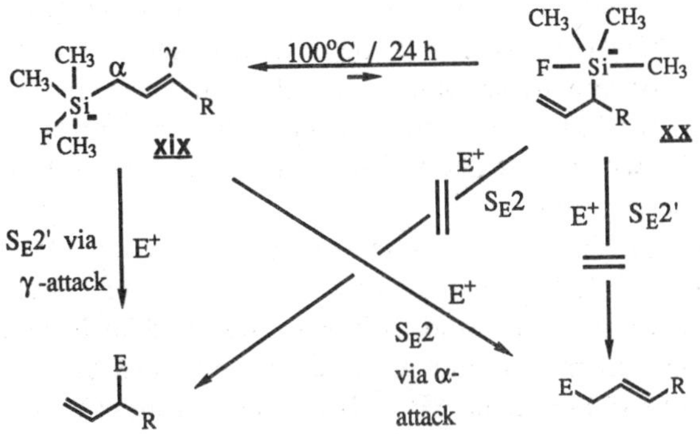

Scheme 58

Organosilicon reactions often proceed via an S_E2 pathway because the long silicon–carbon bond permits electrophiles access to the nucleophilic carbon. Although intriguing, the possibility that silicate **xix** isomerizes to **xx** under fluoride ion catalysis prior to carbon–carbon bond formation can be discounted. Fluoride ion-promoted allylations generally take place with ease at room temperature. Although the equilibration of substituted allylsilanes is known, the conditions required are relatively vigorous (cat. TBAF, THF, 100°C, sealed tube, 24 h), and the more substituted olefin is formed, e.g., **165** → **166** (Scheme 59).[101] These considerations obviate an isomerization of the silicate intermediates.

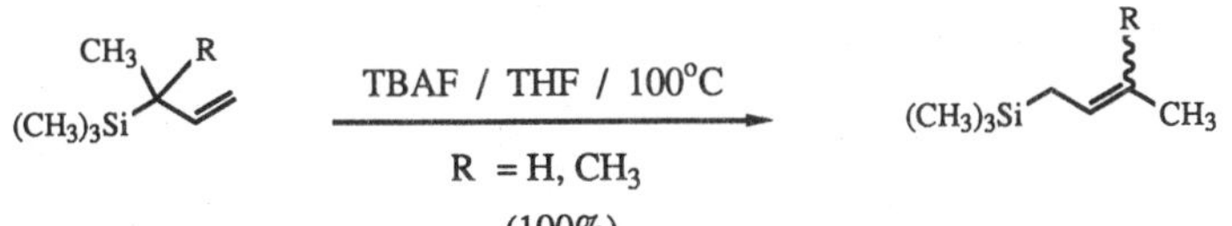

Scheme 59[101]

The use of allylsilanes containing functionality capable of stabilizing a carbanionic intermediate has been briefly explored. Ricci and co-workers observed that fluoride-induced desilylation of cinnamylsilane (**167**) with benzaldehyde gave only the α-alkylated product (Scheme 60).[102] Clearly the styrenyl unit governs the regioselectivity of this reaction. The use of benzyl bromide as an electrophile is the only reported example of the allylation of an alkyl halide under fluoride ion catalysis.[102,103]

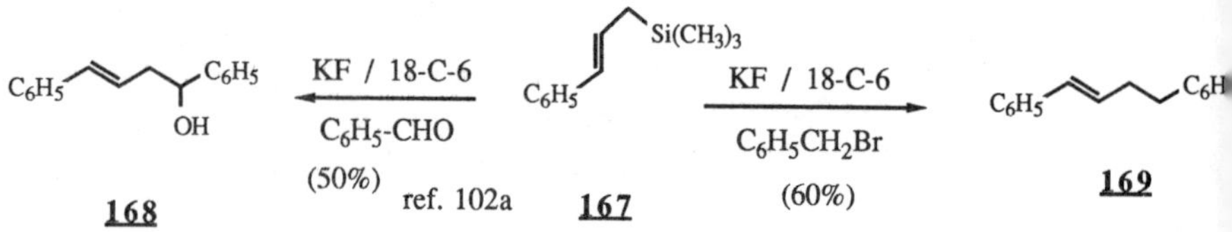

Scheme 60

To date only simple allylsilanes, such as allyltrimethylsilane, have been used in the fluoride-induced allylation procedure. Notable exceptions to this statement are the reaction of 2-[trimethylsilyl)methyl]-1,3-butadiene (**170**),[104] *E*-1,4-bis(trimethylsilyl)-2-butene (**172**),[105] and silylcyclopropane (**174**)[106] with carbonyl compounds (Scheme 61). The pentadienyl anion formed when allylsilane **174** is treated with TBAF also reacts regiospecifically with phenyl cyanate to give nitrile **176**, a key intermediate in Paquette's syntheses of α-vetispirene, hinesol, and β-vetivone.[106]

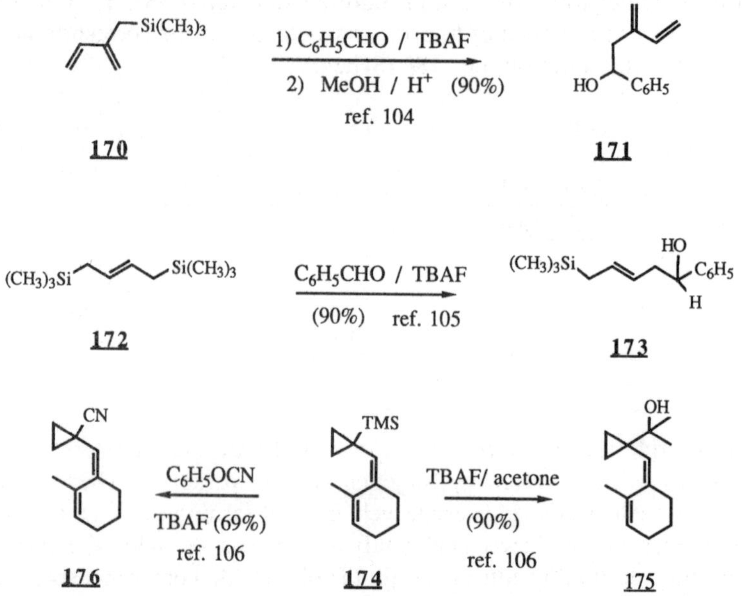

Scheme 61

2. 1,4-Additions

A pivotal result is seen in the 1,4-allylation of *E*-1-phenyl-1-buten-3-one (**102**) by Hosomi and Sakurai (Scheme 62).[92] This allylation was found to generate products resulting from conjugate addition (**177**) in 24% yield and 1,2-addition (**178**) in 50% yield. This lack of selectivity contrasts with

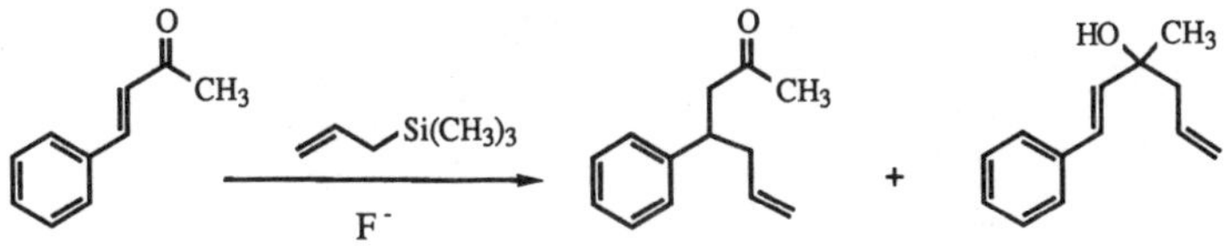

102   **177**   **178**

                 24%   50%

Scheme 62[92]

the regiospecific conjugate allylation of enones activated by Lewis acids, and at first diminished the synthetic utility of the fluoride ion-catalyzed allylation procedure.

The fluoride ion-catalyzed 1,4-allylation of cyclohexenone was achieved by Ricci and co-workers. In their 1982 communication, addition of allyltrimethylsilane, as well as benzyl-, heteroarylsilanes, was described to be exclusively conjugative with cesium fluoride as catalyst (c.f. **179**), whereas use of TBAF as the source of fluoride ion yielded adducts corresponding to both 1,4- and 1,2-addition, i.e., **180** (Scheme 63).[107]

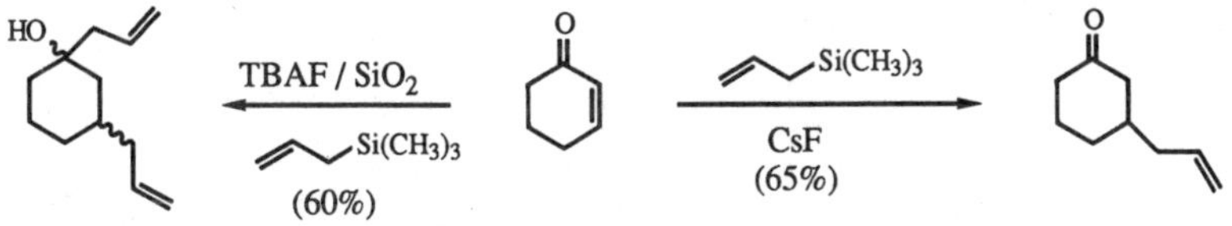

180   **179**

Scheme 63[107]

The fluoride ion-catalyzed allylation of α,β-unsaturated esters, nitriles, and amides was discovered by Majetich and co-workers.[68] They reported that allyltrimethylsilane adds exclusively in a 1,4-fashion to these functional groups, whereas the Lewis acid-catalyzed process (a Sakurai allylation) yielded no addition products at all (Table 2). A comprehensive study revealed that this allylation procedure is also more general than traditional procedures, such as the use of organocuprates. Studies of other unsaturated ketones corroborated Hosomi's and Sakurai's observation that allylation of these substrates is not regiospecific; instead, a marked substrate dependency was found.

The fluoride ion/allyltrimethylsilane procedure also afforded exclusively the 1,4-adduct in allylation reactions with polyene ester **181** and nitrile **182**,[68] in contrast to cuprates, which preferred 1,6-conjugate addition (Scheme 64).[108]

Table 2[68]

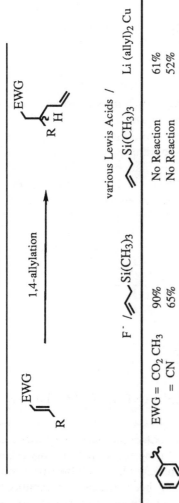

		F^- / $\diagdown$Si(CH$_3$)$_3$	various Lewis Acids / $\diagdown$Si(CH$_3$)$_3$	Li (allyl)$_2$ Cu
R = (phenyl)	EWG = CO$_2$ CH$_3$	90%	No Reaction	61%
	= CN	65%	No Reaction	52%
	= CO-NEt$_2$	80%	No Reaction	No 1,4-addition
R = CO$_2$CH$_3$	EWG = CO$_2$CH$_3$	80%	No Reaction	64%
R = (furyl)	EWG = CO$_2$Et	83%	No Reaction	26%
	= CN	91%	No Reaction	73%
R = (tert-butyl)	EWG = CO$_2$Et	80%	No Reaction	No 1,4-addition
	= CN	65%	No Reaction	47%
R = H	EWG = CO$_2$CH$_2$C$_6$H$_5$	65%	Decomposed	28%

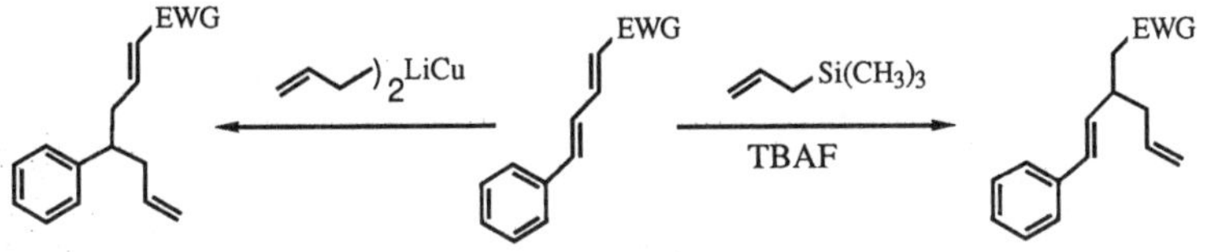

EWG

$)_2^{LiCu}$

EWG

$Si(CH_3)_3$

TBAF

EWG

95% ----- EWG = CO_2Et **181** ---------------- 63%
80% ---------- = CN **182** ---------------- 27%

Scheme 64[68]

Functionalized allylsilanes, such as 1-acyloxy-2-propenyltrimethylsilane (**183**), also react in 1,4-fashion with α,β-unsaturated ketones under fluoride ion catalysis (Scheme 65).[109]

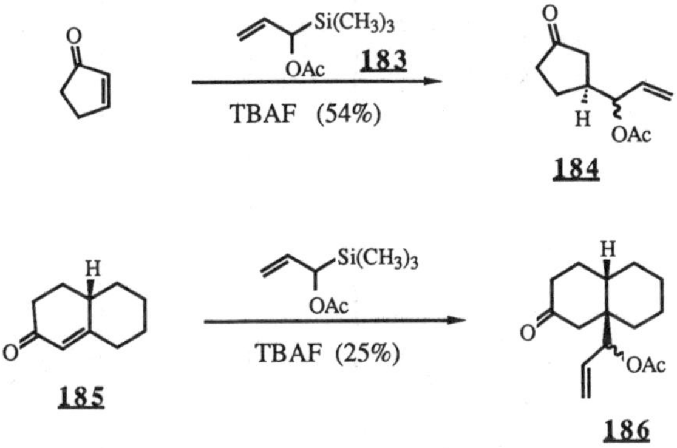

$Si(CH_3)_3$

OAc **183**

TBAF (54%)

184

$Si(CH_3)_3$

OAc

TBAF (25%)

185

186

Scheme 65[109]

III. INTRAMOLECULAR ALLYLSILANE CYCLIZATIONS

A major concern in the total synthesis of polycyclic natural products is annulation, the process of forming and/or fusing rings onto existing cyclic structures. Various electrophilic substitution reactions of allylsilanes, which were found to occur intermolecularly, have been achieved intramolecularly.[110] As before, the following reactions are divided into either electrophilic substitutions or fluoride ion-initiated cyclizations.

A. Electrophilic Substitution Cyclizations

Novel and efficient annulation procedures have been achieved via the Lewis acid-catalyzed intramolecular reaction of allylsilanes with various carbon and heteroatom electrophiles. However, this section begins with intramolecular studies using protons, the simplest electrophiles.

1. Protic Acids

In 1983, Wilson and Price uncovered a novel means of 1,3-asymmetric induction (Scheme 66).[111] Though the esterification of allylsilane **187** with methanolic HCl proceeds without protodesilylation, the use of standard protodesilylation conditions (boron trifluoride and acidic acid) gives (2S,4R)-dimethyl-5-hexenoic acid (**188**) in high selectivity (>8:1). Intramolecular protonation of the allylsilane moiety via a cyclic transition state such as **xxi**, accounts for this selectivity.

Scheme 66[111]

Wilson also studied the effectiveness of other functional groups as the proton source of 1,3- and 1,4-asymmetric inductions (Scheme 67). While the protodesilylation of alcohol **189** took place with the predicted acyclic stereoselection, the protodesilylation of acetate **190** gave a mixture of products in which the isomer with the inverse stereochemistry at the new asymmetric center predominated.

Scheme 67[111]

2. Carbon Electrophiles

Although allylsilanes undergo protodesilylation by the action of protic acids, the high rate of intramolecular reactions permits the use of such acids to generate numerous electrophiles. Nevertheless, most intramolecular electrophilic substitution reactions are catalyzed by common Lewis acids.

a. Carbocations. Bridgehead carbocations derived from small or medium-sized ring systems generally undergo electrophilic substitutions. Kraus and Hon have used this aptitude to prepare functionalized bicyclo[2.2.2]octanes and bicyclo[3.3.1]nonanes.[112] In particular, the bridgehead carbocation generated from halide **191** with silver triflate reacted with the juxtaposed allylsilane unit to afford compound **192** in 75% yield (Scheme 68). Note that this substitution reaction proceeds with retention of configuration because of the constraints imposed by the bicyclic system.

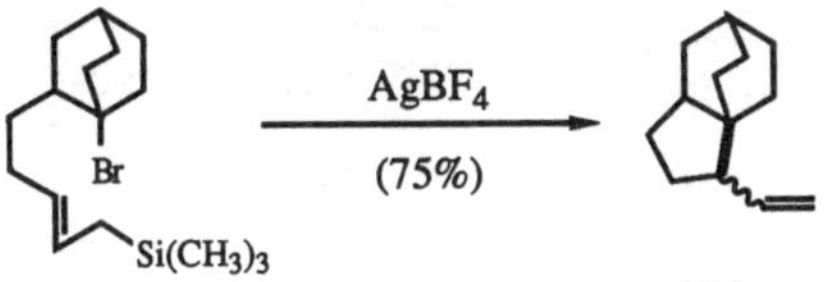

Scheme 68[112]

Allylic cations have been used to initiate intramolecular reactions with allylsilanes. Scheme 69 illustrates the Lewis acid-promoted conversion of allylic alcohols **193** and **195** to bicyclo[3.3.0]octane derivative **194** and tricyclic adduct **196**, respectively.[113] These cyclizations proceed via the following sequence: (1) formation of an allylic cation, (2) addition of this cation to a neighboring double bond, and (3) trapping of the newly formed carbocation by the allylsilane unit present. The use of a 1:1 mixture of titanium tetrachloride and *N*-methylaniline to initiate this reaction is noteworthy since the exocyclic double bond of related systems is prone to isomerize in the presence of Lewis acids.

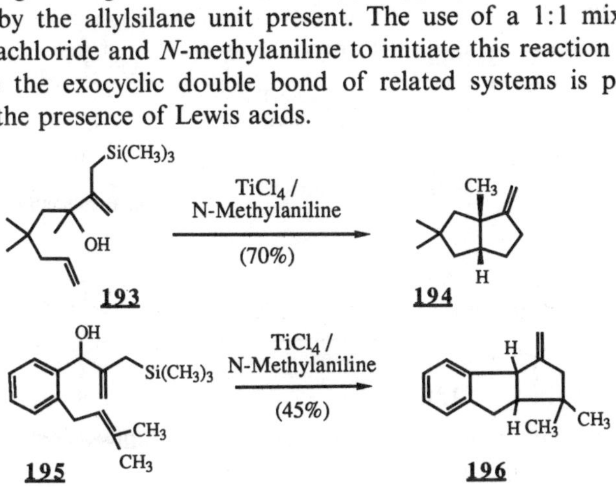

Scheme 69[113]

In 1986, Schreiber and co-workers reported the Lewis acid-mediated version of the Nicholas reaction[114a] for the preparation of six-, seven-, or eight-membered rings (Scheme 70).[114b] They found that treatment of cobalt-complex **197** with boron trifluoride etherate leads to a new intraannular cobalt-complex that can react further, via the Pauson–Khand annelation procedure, to produce polycyclic systems (cf. **198** and **199**).

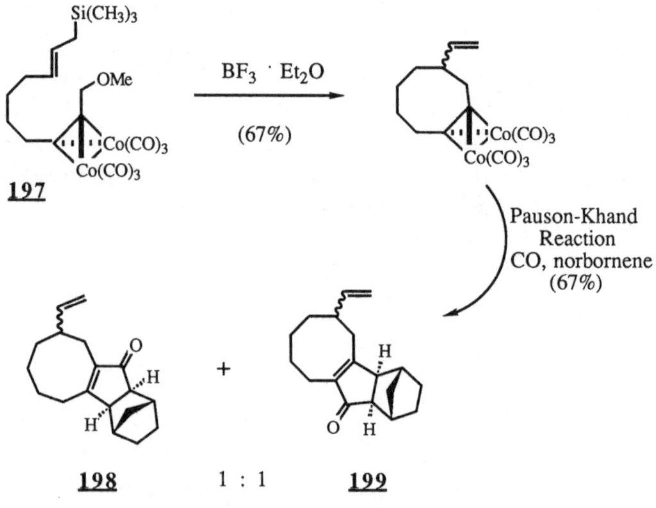

Scheme 70[114]

Johnson and co-workers have demonstrated that allylsilanes are effective terminators of polyene cyclizations.[115] In the biomimetic ring closure of alcohol **200**, capture of the transient tricyclic carbonium ion by the allyl-silane competes favorably with processes such as hydrogen abstraction and anionic capture (Scheme 71).

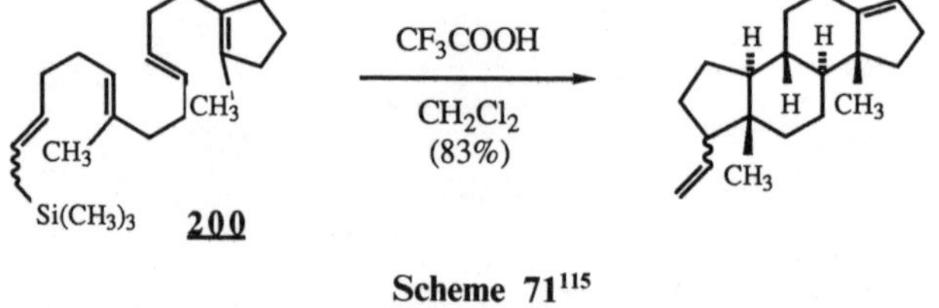

Scheme 71[115]

Weiler and co-workers have prepared functionalized cyclohexanes, such as **201**, by cyclizing functionalized allylsilanes with trisubstituted olefins.[116a]

Further work focused on multiple ring closures of related trienes and culminated in syntheses of albicanyl acetate and isodrimenin (**203**)[116b] (Scheme 72).

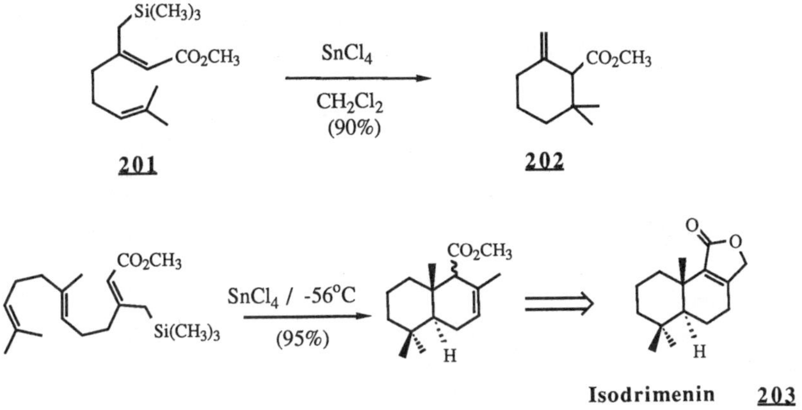

Scheme 72[116]

b. Acetals and Ketals. In 1976, Fleming, Pearce, and Snowden reported the first use of an allylsilane in an intramolecular electrophilic cyclization.[117a] In this pioneering example, the allylsilane moiety controls the outcome of the intramolecular ring closure of dimethyl acetal **204** (Scheme 73). In the absence of the silyl group, five isomeric products are formed. The use of a trimethylsilyl group to dictate the formation of a specific olefin, as is exemplified by the cyclization of allylsilanes **204** and **206**,[117b] represents one of the most important attributes of allylsilane cyclizations.

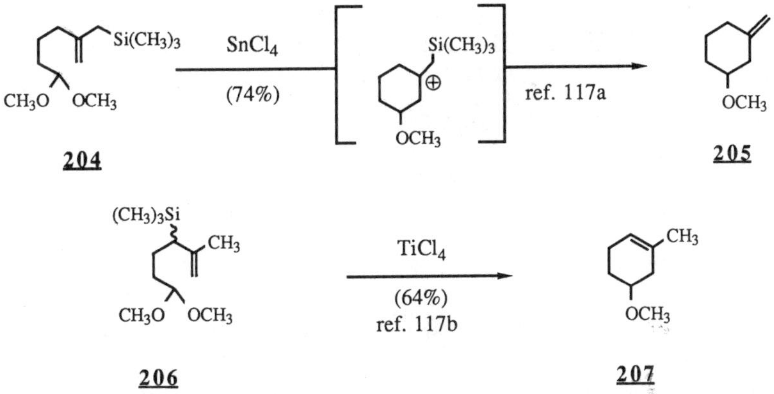

Scheme 73[117]

A more demanding variation of the above cyclizations is shown in Scheme 74. Here, multiple ring closures are initiated by the carbonium ion formed from the acetal (208) and eventually trapped by the allylsilane present.[118]

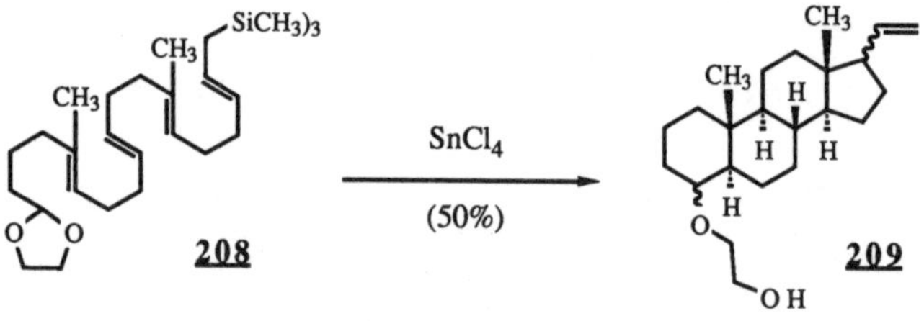

Scheme 74[118]

An elegant, one pot cyclopentane annulation has been reported by Lee and co-workers.[119] Their novel annulation strategy features the use of silane 210 in which an acetal functionality is present (Scheme 75). The coupling of 210 with silyl enol ether 211 occurs when catalyzed by trimethylsilyltrifluoromethanesulfonate (TMSOTf). In addition, under the reaction conditions the TMSOTf also catalyzes an intramolecular Hosomi reaction of the incipient aldehyde unit. Unfortunately, this process is only modestly regioselective (cf. 212/213 and 214/215).

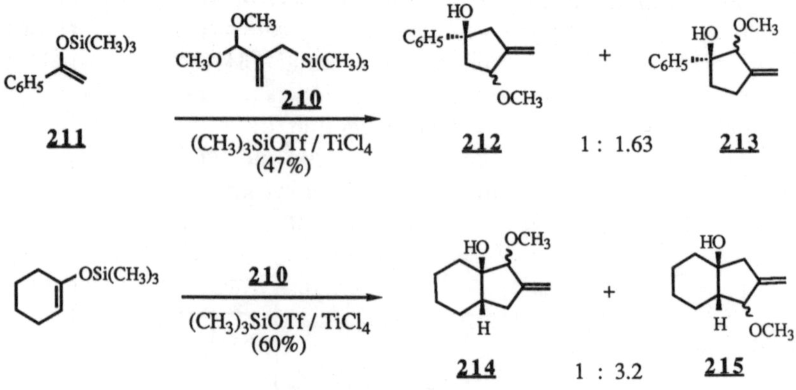

Scheme 75[119]

Chan and co-workers observed that alkoxyallylsilanes condense with carbonyl compounds in the presence of aluminum trichloride to give substituted tetrahydropyrans in good yield (Scheme 76).[120a] This stereoselective condensation was established to proceed through a mixed silyl acetal intermediate. In general, aliphatic aldehydes gave the best yields of disubstituted 4-chlorotetrahydropyrans. Ricci and co-workers have reported a related stereoselective synthesis of oxacycles via acetal-initiated cyclizations (cf. 218 → 219).[120b]

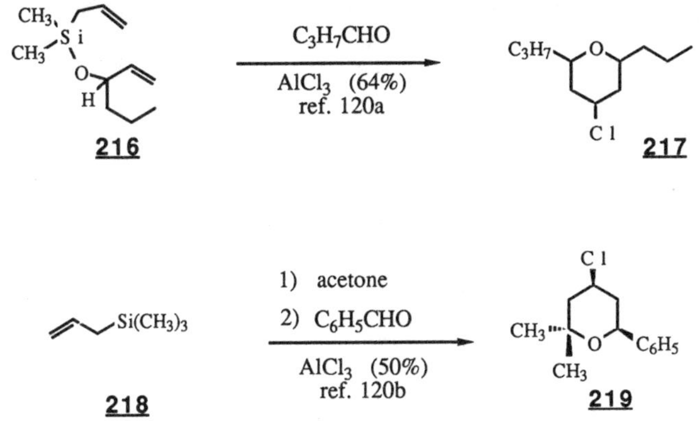

Scheme 76

c. Ketones and Aldehydes. Intramolecular Hosomi Reactions: Chart 5 contains three intramolecular variations of the Hosomi reaction with aldehydes. Moreover, each of these reactions proceeds with significant stereoselectivity.[121] The cyclization of **220** is particularly intriguing, because the allylsilane moiety is chiral and forms cyclopentenol **221** with remarkable enantioselectivity (>98% e.e.).[122] In 1979, Oshima and co-workers[123] reported the annulation of aldehyde **222** to yield vinyl sulfide **223**. This study was one of the first to address the diastereoselectivity of allylsilane cyclizations. Finally, the cyclizations of functionalized allylsilanes to aldehydes have proven useful for the preparation of bicyclic α-methylene lactones (cf. **226**).[124]

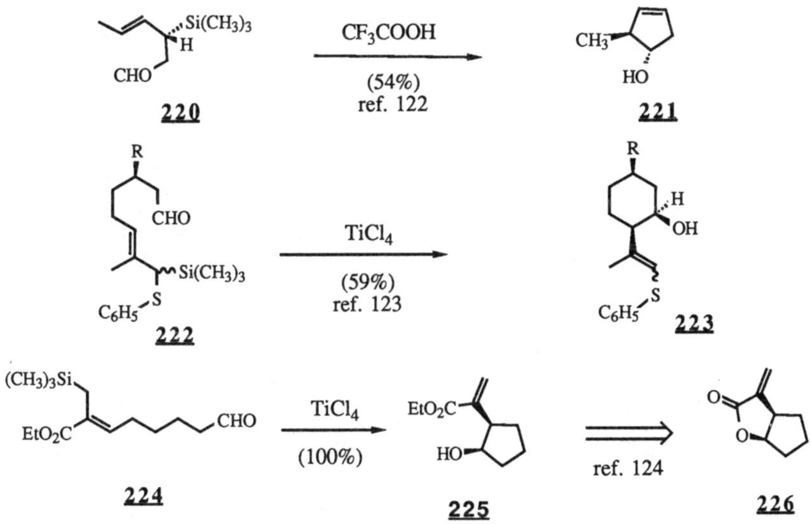

Chart 5

The stereochemical outcome of intramolecular 1,2-additions where the orientation of the reactive centers prior to cyclization is predictable (cf. **227**) was studied by Denmark and Weber.[125] Scrutiny of Scheme 77 reveals a correlation between the covalent radius of the metal used in catalysis and the

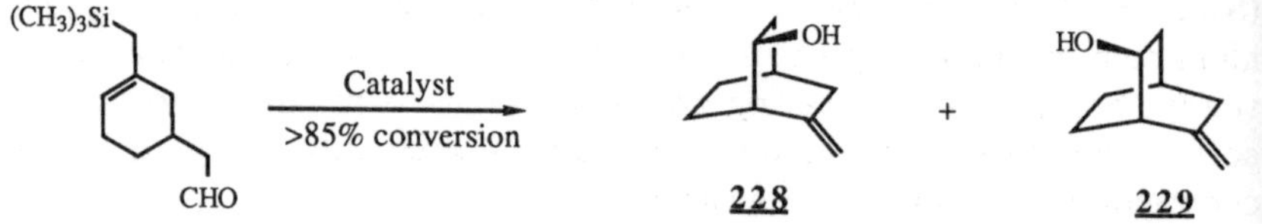

227	Catalyst	**228**	**229**
	$SnCl_4$	49%	51%
	Et_2AlCl	66%	34%
	$FeCl_3$	70%	30%
	$AlCl_3$	79%	21%
	$BF_3 \cdot Et_2O$	80%	20%

Scheme 77

stereoselectivity of the reaction. For example, tin tetrachloride resulted in the lowest selectivity because of its large size in contrast to the use of boron trifluoride etherate, which produced predominantly alcohol **228**.

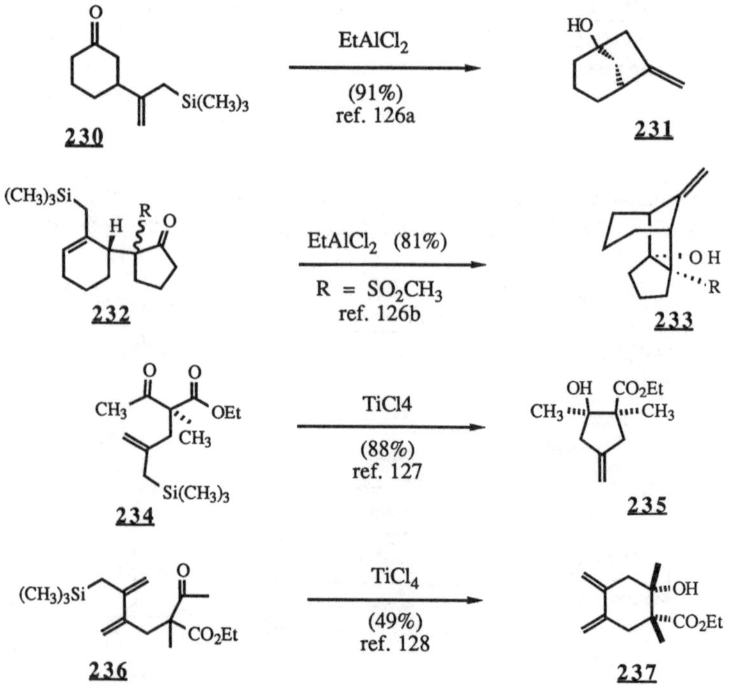

Chart 6

Chart 6 presents four straightforward additions of allylsilanes to ketones.[126-128] Nevertheless, such annulations are often substrate dependent. For example, a rather unusual transformation was observed in the course of the Lewis acid-catalyzed cyclizations of ketones such as **238** (Scheme 78).[129] Although 1,2-addition occurs, in the presence of ethylaluminum dichloride a pinacol-like rearrangement takes place in which the sulfone group acts as a leaving group. This novel spiroannulation process corresponds to the annulation of a five-membered ring, followed by ring contraction of the pre-existing carbocycle.

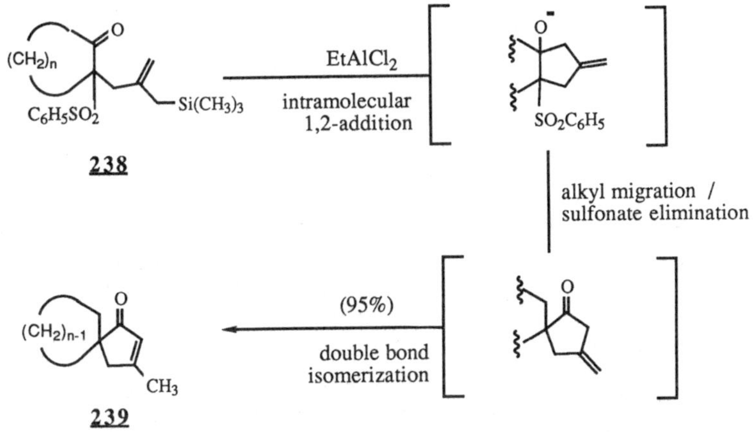

Scheme 78[129]

In the presence of Lewis acids, less electrophilic or sterically congested ketones such as **240** do not necessarily react in 1,2-fashion (Scheme 79).[130c]

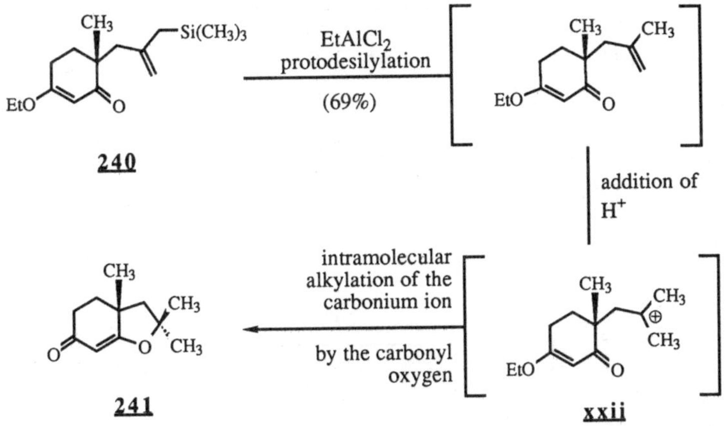

Scheme 79

Instead, protodesilylation of the allylsilane occurs, followed by subsequent addition of HCl to the nascent double bond to produce a tertiary carbonium ion (cf. **xii**), which is then trapped by the carbonyl oxygen atom to yield enone **241** in good yield.

d. α.β-Unsaturated Enones and Enals. Intramolecular Sakurai Reactions: **Wilson and Price** reported that cyclohexanone **243** was produced in 73% yield when allylsilane **242** was treated with boron trifluoride etherate (Scheme 80).[131] This cyclization was the first example of an intramolecular Sakurai reaction.

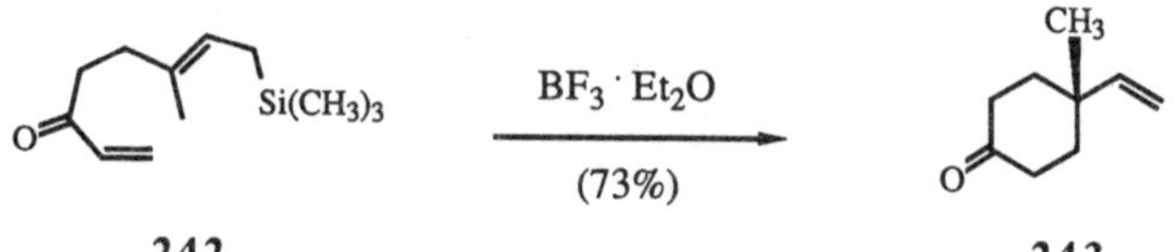

Scheme 80[131]

Other research groups soon reported their independent examples of intramolecular addition of allylsilanes to various Michael acceptors. Chart 7 presents eight substrates reported by Majetich and co-workers illustrating the versatility and some of the limitations of this annulation procedure.[132] In general, conjugate addition predominates if the allylsilane can easily adopt a spatial disposition favorable for 1,4-attack; otherwise, 1,2-addition or protodesilylation predominates. Nitrogen-containing substituents complex with the catalyst leading to protodesilylation. Note that the reactions of entries **246** and **247** proceed with useful diastereoselectivity.[121,133]

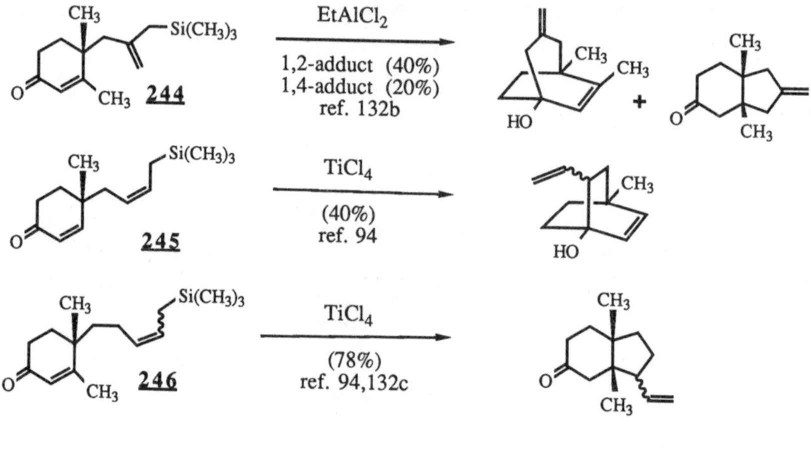

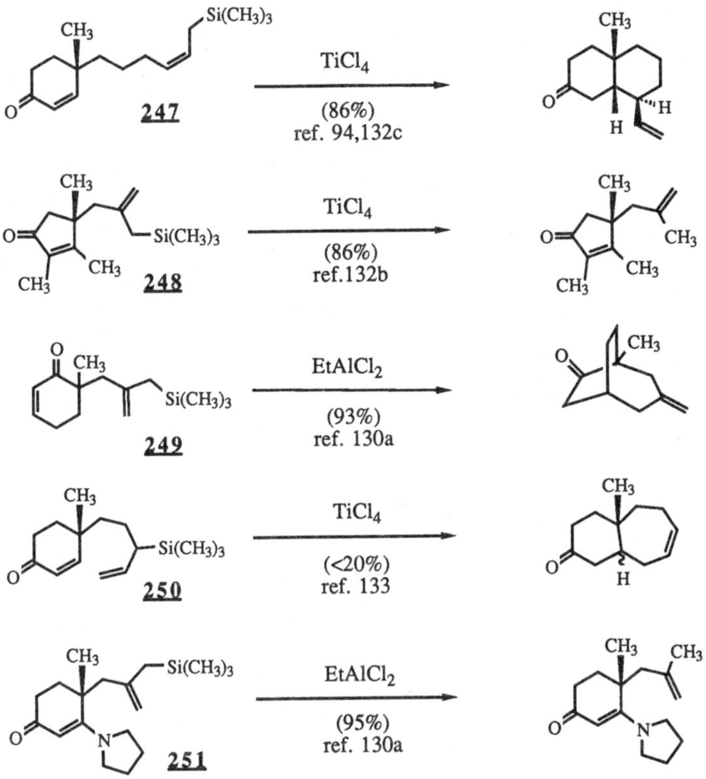

Chart 7

To date there are few examples of the application of intramolecular addition of allylsilanes to enones in natural product synthesis. The reactions of enone **252** presented in Scheme 81 have established that tricyclic systems can be assembled with this methodology.[134] This ability to generate polycyclic systems will undoubtedly facilitate the design of synthetic routes to complex natural products.

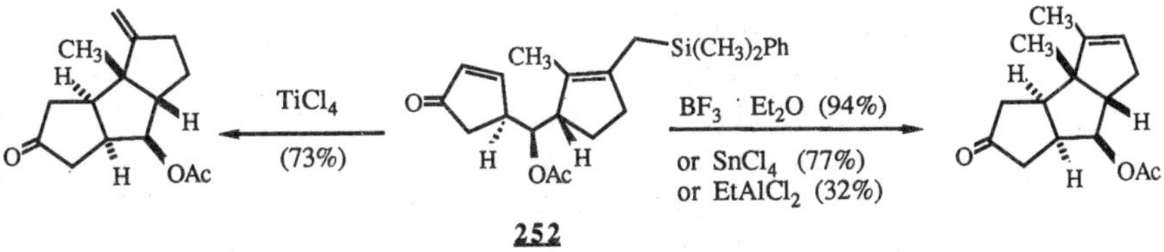

Scheme 81[134]

Others have also recognized the synthetic potential of intramolecular allylsilane additions to electron-deficient olefins (Chart 8). Schinzer and co-workers have used this methodology to prepare functionalized spiroketones (254)[135a] and substituted hydrindanones (256).[135b] Tokoroyama and co-workers observed that cyclization of allylsilane 257 with $TiCl_4$ gave decalones 258 and 259 in 90% yield, as a 3:2 mixture of *trans*- and *cis*-fused isomers.[136] Equilibration of 259 to 258 revealed that this reaction occurs with remarkable stereoselectivity.

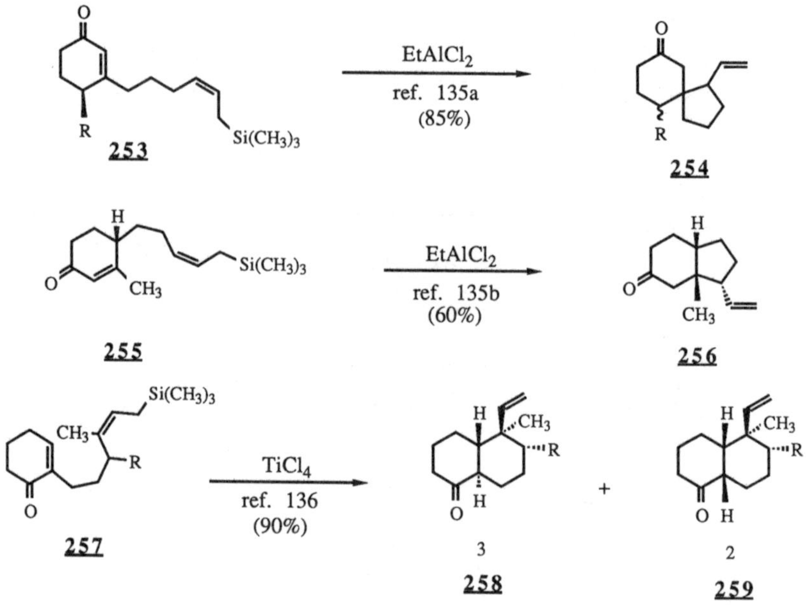

Chart 8

e. Dienones. The use of dienones as electrophiles represented a new approach to carbocyclizations. Majetich and co-workers pioneered and have developed this novel annulation strategy, which has proved extremely useful for the preparation of medium-sized rings (Chart 9).[137-139] For example, entries 260, 262, 265 and 267 annulate cycloheptane rings in good yield, while the cyclizations of dienones 263 and 264 give bicyclic cyclooctanes. Decalones and functionalized tricycles can also be prepared by this methodology in very good yield. The cyclizations of 261, 262, 263, and 267 occur with stereospecificity.[140] Scheme 82 shows the key transformations in Majetich's syntheses of nootkatone (270),[137c] perforenone (271),[78,137d] and *epi*-widdrol (272)[137e] that feature this carbocyclization strategy. In 1986 this annulation approach was confirmed by Schinzer and co-workers (Scheme 83).[141]

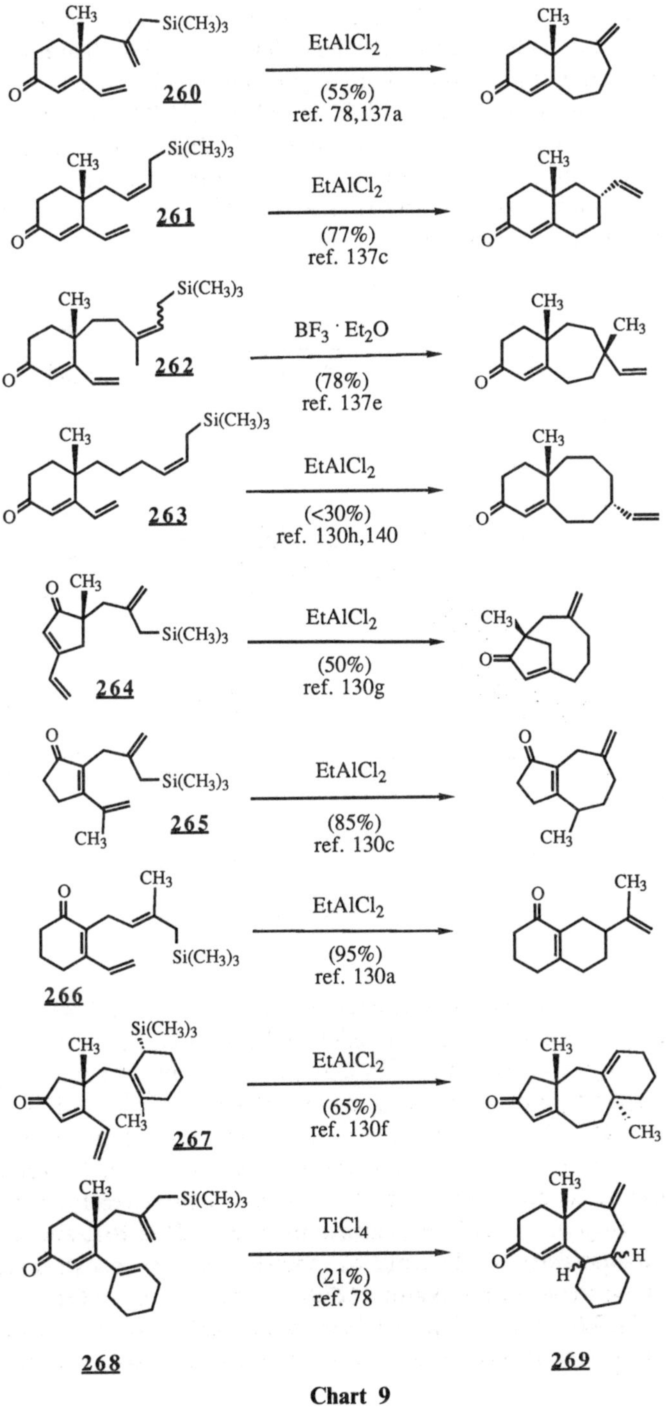
CH3
Si(CH3)3
260
EtAlCl2
(55%)
ref. 78,137a
261
EtAlCl2
(77%)
ref. 137c
262
BF3 · Et2O
(78%)
ref. 137e
263
EtAlCl2
(<30%)
ref. 130h,140
264
EtAlCl2
(50%)
ref. 130g
265
EtAlCl2
(85%)
ref. 130c
266
EtAlCl2
(95%)
ref. 130a
267
EtAlCl2
(65%)
ref. 130f
268
TiCl4
(21%)
ref. 78
269

Chart 9

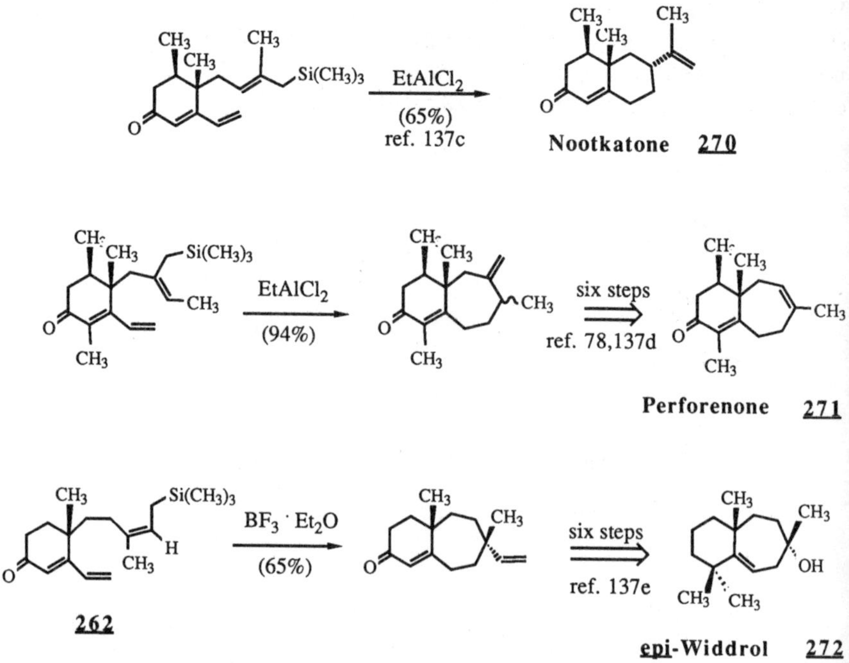

Nootkatone **270**

Perforenone **271**

262

epi-Widdrol **272**

Scheme 82

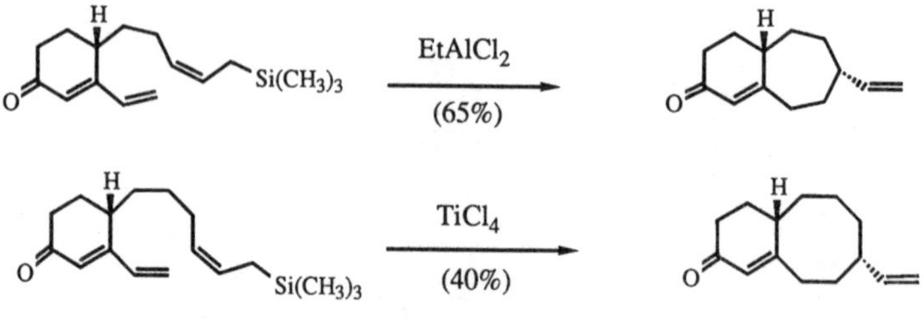

Scheme 83[141]

3. *Other Electrophiles*

Epoxides cyclize in the presence of Lewis acids or methanolic HCl to construct five-, six-, or seven-membered rings. This strategy was used to prepare cyclopentanols **274** and **275**, in which the *cis* product predominates (Chart 10).[142] The formation of cycloheptanol **276** by Wang and Chan represents a novel approach for the construction of seven-membered rings.[143] The cyclization of epoxy silane **277** was featured by Armstrong and

Weiler in an elegant synthesis of karahana ether **278**.[144a] These workers also studied the cyclization of allylsilane **279**.[144b] Finally, Cutting and Parsons noted an unexpected epoxide rearrangement, rather than intramolecular epoxide opening, on treatment of **280** with stannic chloride.[145]

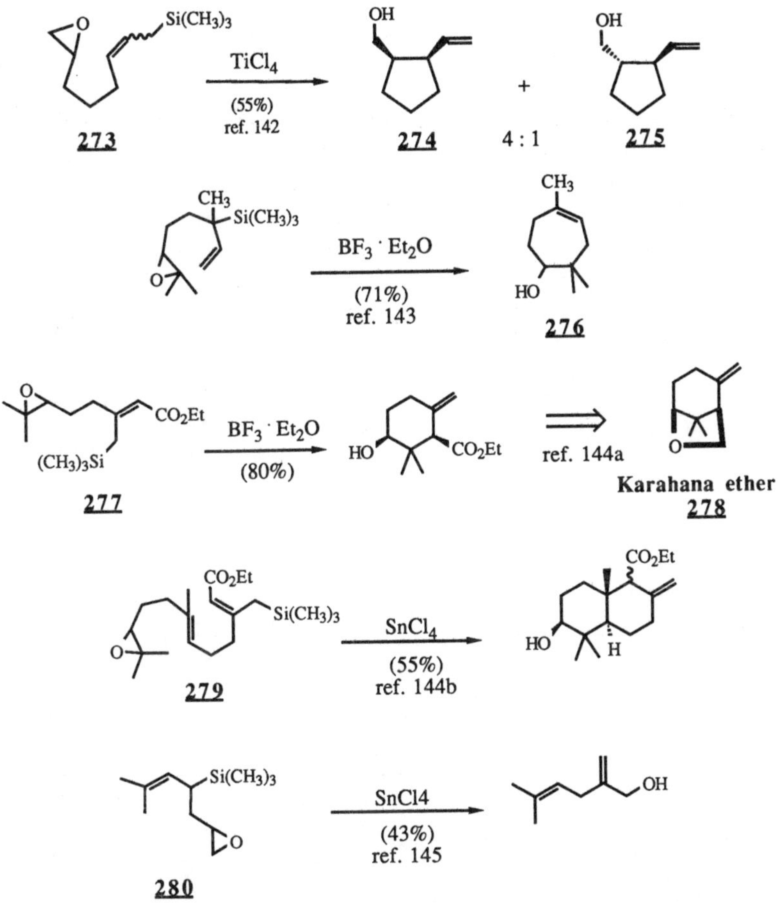

Chart 10

Akiba and co-workers have developed a new approach to the synthesis of medium-sized lactones via Lewis acid-promoted addition of α-chlorosulfides and a suitably tethered allylsilane moiety (Scheme 84).[146] This methodology was showcased in a direct synthesis of phoracantholide (**281**).

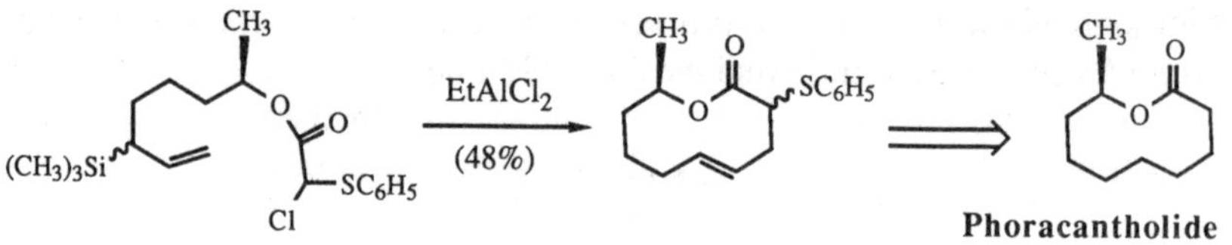

Phoracantholide

281

Scheme 84[146]

The work of Hiemstra and Speckamp has established the usefulness of *N*-acyliminium ions in organic synthesis.[147] These scientists have developed several ways to prepare cyclic *N*-acyliminium ions and have conducted a systematic study of their reactivity with allylsilanes and propargylsilanes. Chart 11 summarizes the application of this methodology to the synthesis of functionalized piperidines as well as isoretronecanol **282**. A synthesis of mesembrine (**283**) was achieved by Gramain and Remuson via an analogous α-acyliminium ion cyclization.[148]

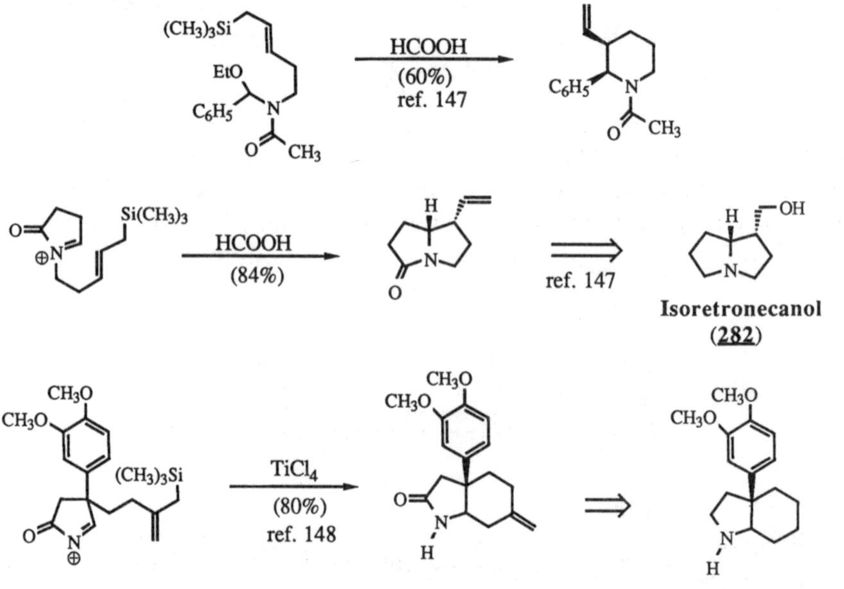

Isoretronecanol
(**282**)

Mesembrine (**283**)

Chart 11

The intramolecular aminomethano desilylation process developed by Grieco and Fobare permits the formation of six-, seven-, and eight-membered nitrogen heterocycles. The key reaction in a direct synthesis of yohimbone (**284**) (Chart 12) utilized a concerted iminium ion-induced polyolefin cyclization terminated by an allylsilane.[149]

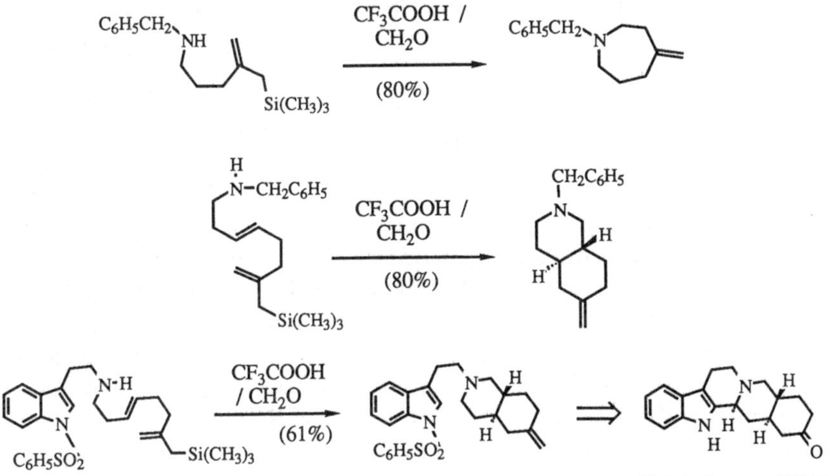

Yohimbone (284)

Chart 12[149]

Allylsilanes react intramolecularly with iminium salts upon photolysis to give homoallyl amines. Mariano and co-workers have probed the scope and generality of this photocyclization methodology (Scheme 85).[150] Their efforts have resulted in the efficient synthesis of various erythrinane derivatives (cf. **286**).

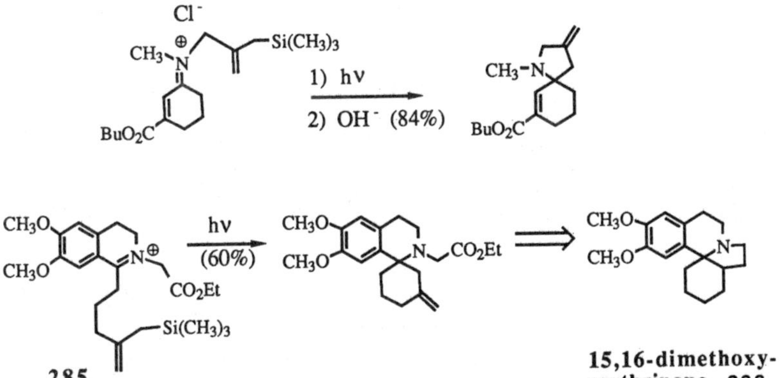

285

15,16-dimethoxy-erythrinane 286

Scheme 85[150]

The iodine-mediated cyclization of allylsilanes **287** and **288** resulted in the formation of five- or six-membered ring ethers or lactones, respectively (Scheme 86).[151] These transformations were promoted by a hypervalent organoiodine compound and a Lewis acid.

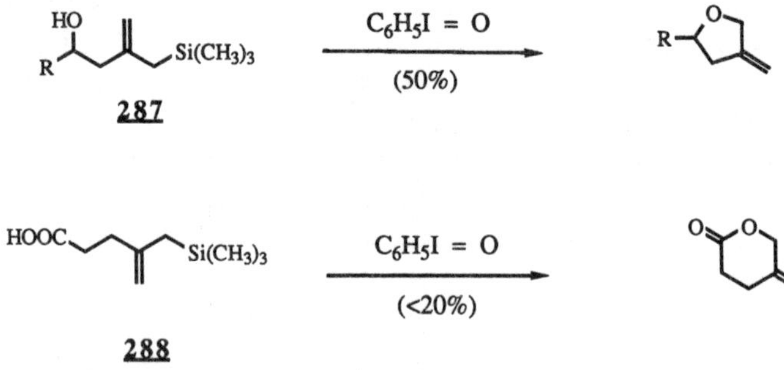

Scheme 86[151]

Lowery[130e] and Majetich observed that the cyclopropylcarbinyl cation generated by treating cyclopropyl ketone **289** with titanium tetrachloride reacts with the allylsilane moiety present to produce perhydroazulene **290** (Scheme 87). Note that the bond formation results in the expansion of the pre-existing cyclohexane ring. Optimization of this process is under way.

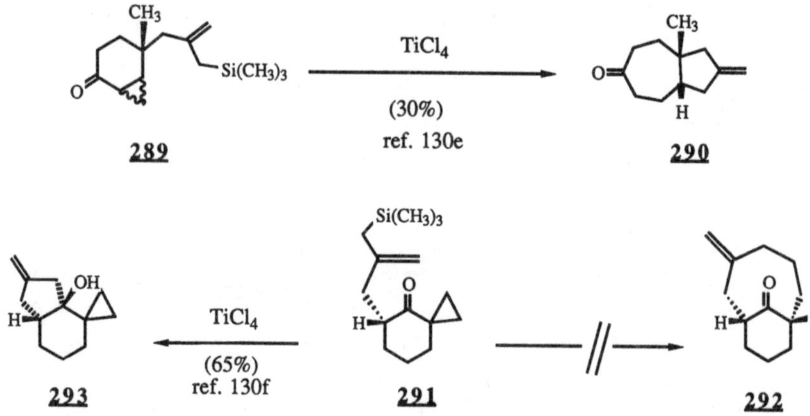

Scheme 87

Opening of cyclopropyl ketone **291** was expected to produce bicyclic ketone **292** containing an eight-membered ring. However, experimentation showed that this cyclization favors the formation of cyclopropylcarbinol **293**.

B. Fluoride Ion-Promoted Allylsilane Cyclizations

1. 1,2-Allylations of Ketones and Aldehydes

As mentioned earlier, Sarkar and Andersen reported the first intramolecular addition of an allylsilane to a carbonyl compound (Scheme 88).[91] Although the reaction of **294** occurred with either Lewis acid or fluoride ion catalysis, the cyclized products consisted of mixtures of axial (**295**) and equatorial (**296**) isomers; however, note the contrasting results of fluoride ion catalysis versus Lewis acid catalysis.[152] More importantly, this study demonstrated that allylsilanes, which previously had been assumed to be labile, could be carried through a large number of routine synthetic transformations.

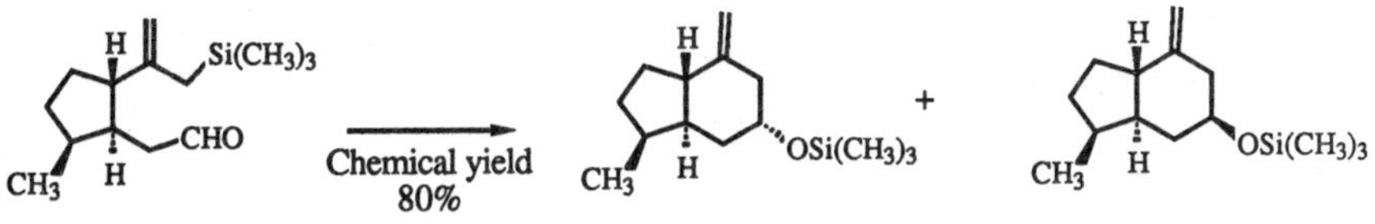

	294	295	296
		85%	15%
$BF_3 \cdot Et_2O$		85%	15%
$SnCl_4$		51%	49%
F^-		18%	82%

Chemical yield 80%

Scheme 88[91]

Trost and Curran soon capitalized on the ability of allylsilanes to add in 1,2-fashion to carbonyl compounds. Treatment of β-ketosulfone **297** with fluoride ion produced *cis*-fused methylenecyclopentane **298**, a key intermediate in Trost's coriolin (**299**) synthesis (Scheme 89).[153]

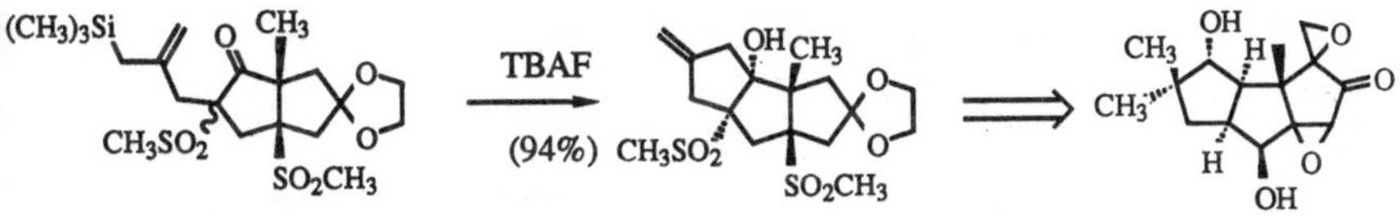

297   **298**   Coriolin
299

TBAF (94%)

Scheme 89[153]

The adduct derived from 1,2-addition of an allylsilane to a β-ketosulfone is also useful for ring expansion. Reaction of cyclopentanol **300** with potassium hydride results in fragmentation with concurrent elimination of benzenesulfinic acid to give cyclooctadienone **301** (Scheme 90).[154] As shown,

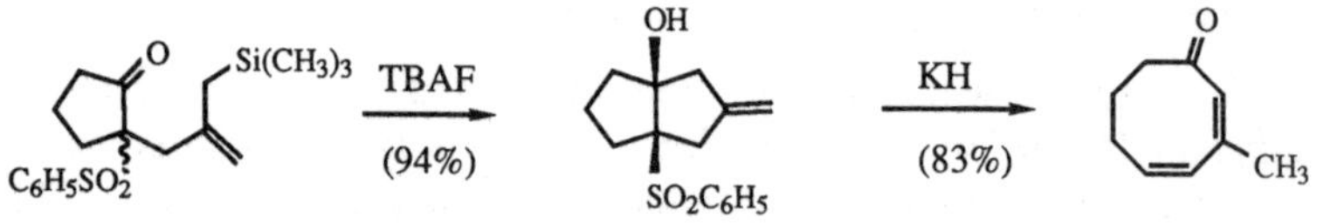

300   **301**

Scheme 90[154]

Trost's three-carbon ring expansion sequence was featured in a four-step synthesis of muscone (**302**) starting with cyclododecanone (62% overall yield, Scheme 91).

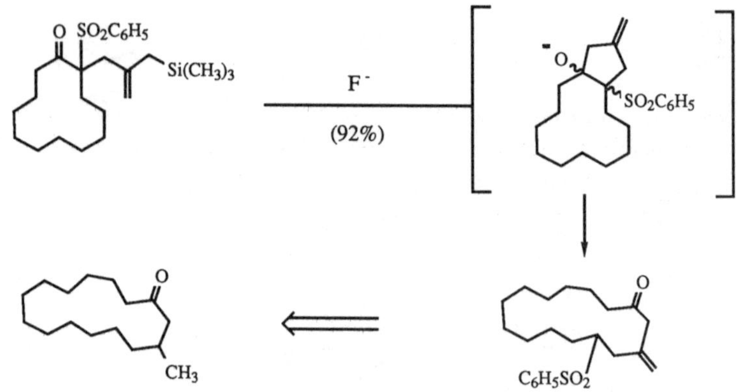

Muscone   **302**

Scheme 91[154]

The fluoride ion-induced cyclization of allylsilane **303** gives 1,2-dimethylenecyclohexane **304** in good yield (Scheme 92).[155a] Hosomi and Sakurai have used this approach to realize other 1,2-dimethylenecyclohexanes and their oxa analogues. Moreover, Trost and co-workers have shown that these compounds are useful cycloaddition reagents.[155b]

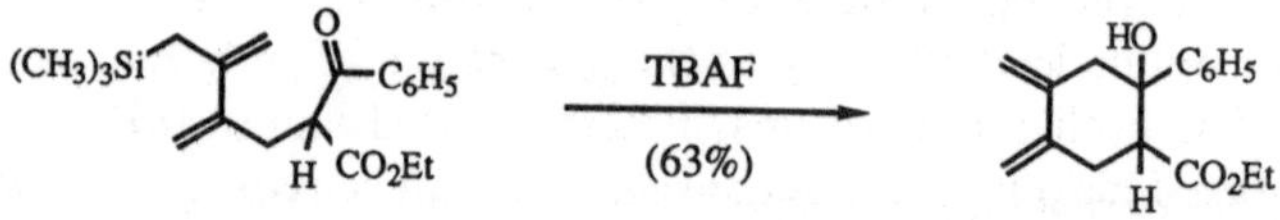

303   **304**

Scheme 92[155]

2. 1,4- and 1,6-Cyclizations to Michael Acceptors

Majetich and co-workers were the first to use fluoride ion-induced conjugate allylation for ring formation. Treatment of unsaturated ester **305** and nitrile **306** with tetra-*n*-butylammonium fluoride gave good yields of annulated products (Scheme 93).[132a,b] These cyclizations proceed despite the presence of both enolizable protons and severe steric interactions, situations presumed to be unfavorable for intramolecular Michael additions.

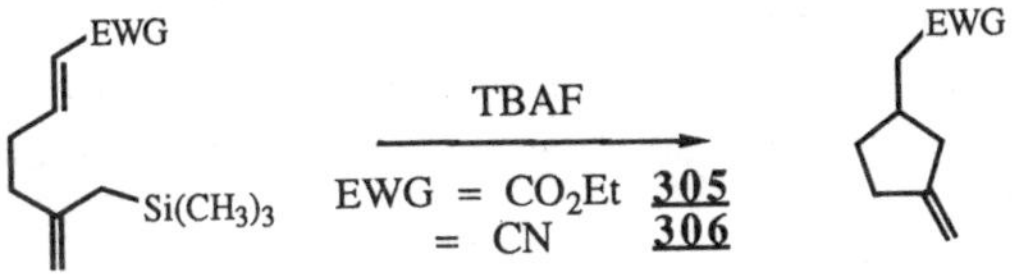

EWG = CO₂Et **305**
     = CN   **306**

Scheme 93[132a]

Fluoride ion-induced cyclizations are effective for the annulation of substituted cyclopentanes onto cyclohexanones or cyclopentenones (Scheme 94).[132a,b] This annulation is capable of generating adjacent quaternary centers. Significantly, Lewis acids gave predominantly 1,2-adducts.

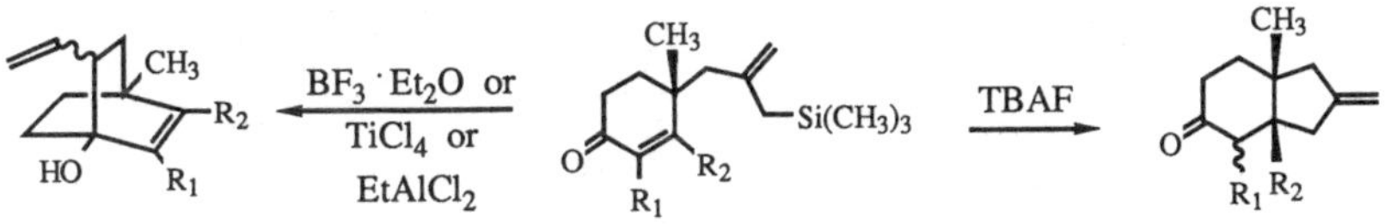

Scheme 94[132b]

A comprehensive investigation of the intramolecular addition of symmetrical and unsymmetrical allylsilanes to simple enones indicated that the regiochemical control exhibited in these ring closures depends solely on kinetic-ring size biases (Chart 13).[156] For example, enone **311** reacts in 1,2-fashion under fluoride catalysis (cf. the cyclization of **249**). To date, annulations based on an addition/elimination strategy have failed (cf. **312**).[130a,b] In addition, the stereoelectronically favored axial approach of the allylic nucleophile leads directly to the generation of only *cis*-fused bicyclic products. The cyclizations of unsymmetrical allylsilanes **314** and **315** produced *cis*-fused bicyclic adducts consisting of two stereoisomers; the ratio of these epimers is controlled by the choice of reaction catalyst. A mechanistic interpretation of these different stereochemical trends (generalized below) has been presented elsewhere.[132c]

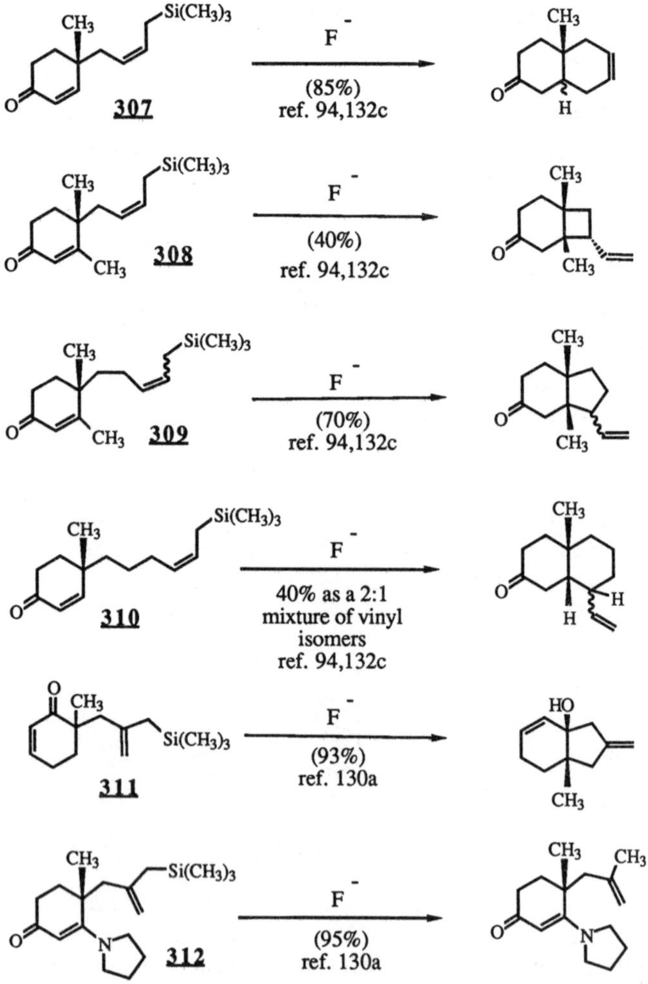

Chart 13

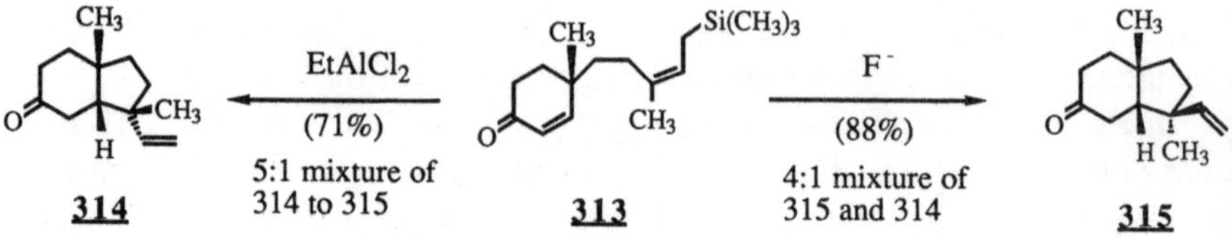

Scheme 95

The use of fluoride ion-initiated conjugate allylations in total synthesis has been limited. Recently, however, Majetich and Defauw have reported that the treatment of enone **316** with fluoride ion generates dienone **xxiii**, which because of geometrical limitations can react only in 1,4-fashion to give tricyclic enone **317** having necessary *cis,anti,cis*-cyclopenta[*a*]pentalene skeleton characteristic of hirsutene (Scheme 96).[134]

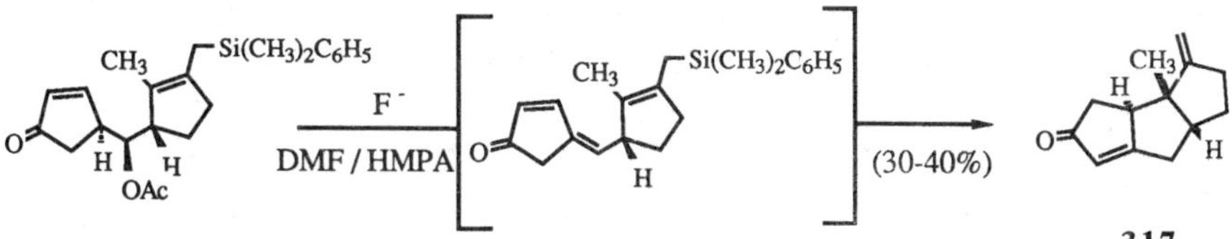

316 xxiii 317

Scheme 96[134]

To date, a major limitation of the fluoride ion-induced cyclizations appears to be their inability to undergo 6-endo-trigonal ring closures.[157] Scheme 97 contains an interesting study whereby allylsilane **320** was pre-

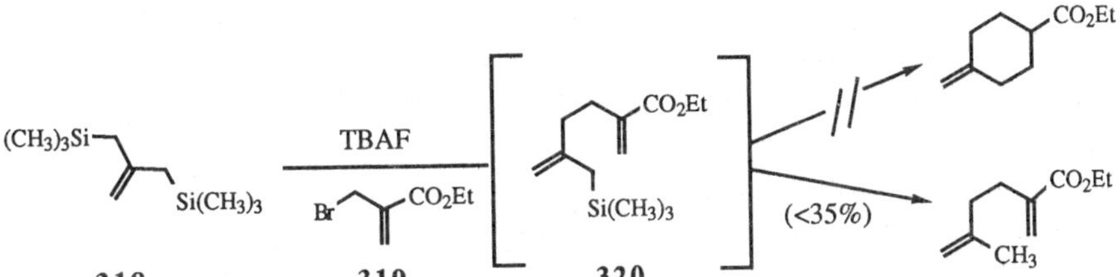

318 319 320 321

Scheme 97[130b]

pared *in situ* by allylation of acrylate **319** with bisallylsilane **318**. Although the intermolecular addition/elimination condensation occurred, the subsequent 6-endo-tricyclization to form a cyclohexane ring failed.[130b]

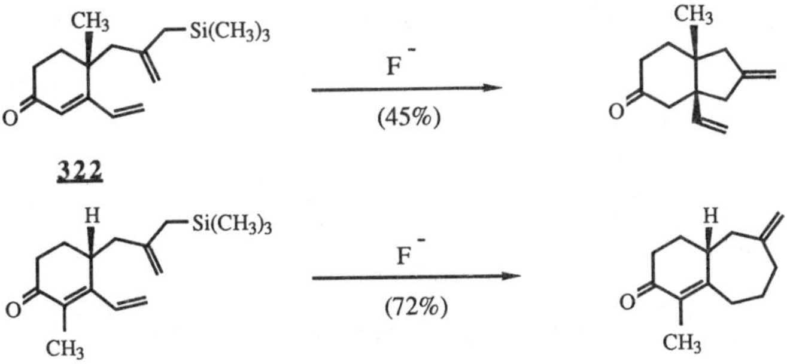

322

323

Scheme 98[78]

The fluoride ion-catalyzed intramolecular addition of allylsilane to dienones has shown that substitution at the γ-position governs whether 1,4- or 1,6-conjugate addition predominates in these cyclizations (cf. **322** and **323**, Scheme 98).[78]

Treatment of substrate **324** with a stoichiometric amount of fluoride ion led to a mixture of bicyclic enone **325** and the bicyclo[2.2.2]octenol derivative **326** in 35% yield and 32% yield, respectively (Scheme 99).[137b] Con-

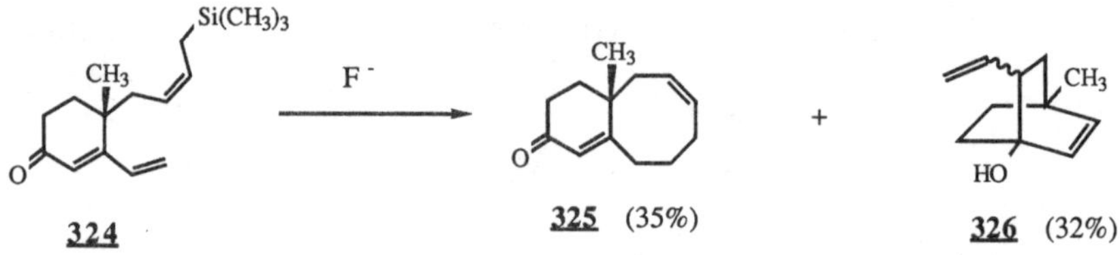

Scheme 99[137b]

ditions have been developed so that cyclization of 2 g of dienone, such as **327**, routinely yields over 50% of the desired cyclooctane product **328**, uncontaminated by the 1,2-adduct. This methodology presents an attractive solution to assembling the carbocyclic skeletons of cyclooctanoid natural products, such as neolemnane (**329**) and precapnellenediene (**330**) (Scheme 100).[130e] Extensive work suggests that these cyclooctane annulations

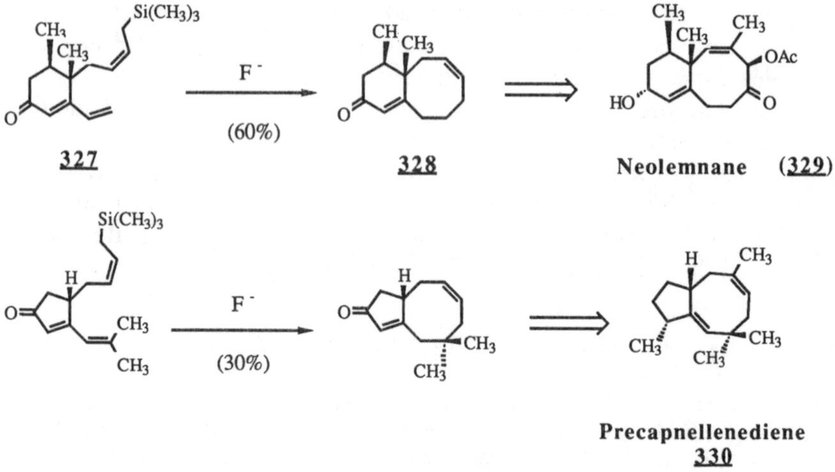

Scheme 100

proceed via a tandem Michael/enolate-accelarated Cope mechanism (**xxv** → **xxvi** → **328**, Scheme 101).[158]

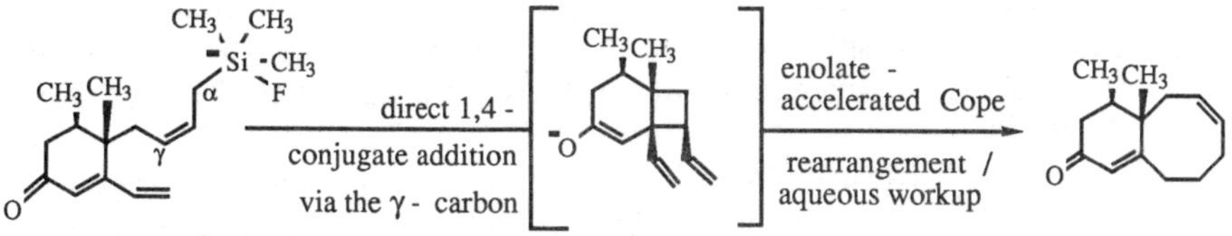

Scheme 101[158]

IV. SUMMARY

As a result of the remarkable versatility and broad synthetic utility demonstrated to date, it is not surprising that reactions of allylsilanes are becoming an increasingly popular method for inter- and intramolecular carbon–carbon bond formation. This area will undoubtedly be the subject of continued interest.

V. EPILOGUE

There is a long standing custom of designating organic reactions with the names of the chemist or chemists who discovered or developed them. In 1983 Blumenkopf and Heathcock published a manuscript entitled "Stereochemistry of the Sakurai Reaction. Additions to Cyclohexenones and Cycloheptenones;"[67a] thus, the "Sakurai Reaction" became part of the chemical literature. However, in declaring the Lewis acid-catalyzed addition of an allylsilane to *Michael acceptors* the "Sakurai Reaction," the co-pioneer of this chemistry was unintentionally overlooked. To correct this oversight, in this review I have taken the liberty of referring to the Lewis acid-catalyzed additions of allylsilanes to *carbonyl compounds* as "The Hosomi Reaction." It is hoped that this suggestion will be accepted.

REFERENCES AND NOTES

1. Confalone and Ko have reported a daring intramolecular nitrile oxide cycloaddition to assemble maytansine's macrocyclic skeleton. See Confalone, P. N.; Ko, S. S. *Tetrahedron Lett.* **1984**, *25*, 947. Thus, hindsight suggests that the reaction illustrated in Scheme 1 warrants reconsideration.
2. (a) Sakurai, H. *Orangosilicon and Bioorganosilicon Chemistry Structure, Bonding, Reactivity and Synthetic Applications,* John Wiley, New York, 1985. (b) Weber, W. P. *Silicon Reagents for Organic Synthesis*, Springer-Verlag, New York, 1983. (c) Colvin, E. W. *Silicon in Organic Synthesis,* R. E. Kreiger Publishing Compnay, Florida, 1981. (d) Larson, G. *Organometallic Chemistry Reviews: Annual Surveys: Silicon–Lead,* Elsevier, New York, 1985, pp. 141–412.

3. General reviews: (a) Fleming, I. *Chem. Soc. Rev.* **1981**, *10*, 83. (b) Fleming, I. In *Comprehensive Organic Chemistry*, Barton, D.; Ollis, W. D., eds., Pergamon Press, Oxford, 1979, Vol. 3, p. 541. (c) Colvin, E. W. *Chem. Soc. Rev.* **1978**, *7*, 15. (d) Hudrlik, P. F. In *New Applications of Organometallic Reagents in Organic Synthesis*, Seyferth, D., ed., Elsevier, Amsterdam, 1976, p. 127.

4. (a) Sakurai, H. *Pure Appl. Chem.* **1982**, *154*, 1. (b) See ref. 2c, pp. 97–124. (c) See ref. 2b, pp. 173–205. (d) Hosomi, A.; Sakurai, H. *J. Synth. Org. Chem. Jpn.* **1985**, *43*, 406.

5. Hosomi and Schinzer have recently published limited surveys of allylsilane chemistry; see Schinzer, D. *Synthesis* **1988**, 263 and Hosomi, A. *Acc. Chem. Res.* **1988**, *21*, 200. Two other comprehensive surveys of this field [by Fleming and Dunogues and Yamamoto] are in press.

6. For exaples of [3 + 2] annulations with allylsilanes, see (a) Trost, B. M. *Angew. Chem. Int. Ed. Engl.* **1986**, *25*, 1–114 and references cited therein. (b) Trost, B. M.; Lynch, J.; Renaut, P. *Tetrahedron Lett.* **1985**, *26*, 6313. For an example of a [4 + 3] annulation, see (c) Hosomi, A.; Otaka, K.; Sakurai, H. *Ibid.* **1986**, *27*, 2881. For an example of a [6 + 3] allylsilane cycloaddition, see (d) Trost, B. M.; Seoane, P. R. *J. Am. Chem. Soc.* **1987**, *109*, 615.

7. Pauling, L. *The Nature of the Chemical Bond*, 3rd ed., Cornell University Press, Ithaca, New York, 1960.

8. *CRC Handbook*, Weast, R. C.; Astie, M. J.; Beyer, W. H., eds., CRC Press, Boca Raton, Florida, 1983, pp. 176–F186.

9. Greene, T. W. *Protective Groups in Organic Synthesis*, Wiley Interscience, New York, 1981, pp. 39–50.

10. For a review on hard/soft acid/base theory, see Saville, B. *Angew. Chem. Int. Ed. Engl.* **1967**, *6*, 928.

11. Ho, T.-L. *Tetrahedron* **1985**, *41*, 1.

12. Nakamura, E.; Kuwajima, I. *Angew. Chem. Int. Ed. Engl.* **1976**, *15*, 498.

13. Holmes, A. B.; Jennings-White, C. L. D.; Schultheis, A. H. *J. Chem. Soc., Chem. Commun.* **1979**, 840.

14. For a comprehensive review of the use of the C–Si(CH$_3$)$_3$ moiety as a protected carbanion see Andersen, N. H.; McCrae, D. A.; Grotjahn, D. B.; Gabhe, S. Y.; Theodore, L. J.; Ippolito, R. M.; Sarkar, T. K. *Tetrahedron* **1981**, *37*, 4079.

15. Burford, C.; Cooke, F.; Ehlinger, E.; Magnus, P. *J. Am. Chem. Soc.* **1977**, *99*, 4536.

16. A reason given for this stabilization is an energy lowering overlap (back bonding) of the empty 3d orbitals on silicon with the adjacent carbon–metal bond. See Huheey, J. E. *Inorganic Chemistry*, 2nd ed., Harper and Row, New York, 1983.

17. The silyl–Wittig reaction is commonly referred to as a Peterson olefination. See Peterson, D. J. *J. Org. Chem.* **1968**, *33*, 780.

18. Sommer, L. H.; Tyler, L. J.; Whitmore, F. C. *J. Am. Chem. Soc.* **1948**, *70*, 2872.

19. Cook, M. A.; Eaborn, C.; Walton, D. R. M. L. *J. Organomet. Chem.* **1970**, *24*, 301.

20. The concept of σ–π conjugation can be invoked to explain the β-effect.[21] This postulate assumes that the electron-rich carbon–silicon bond is able to overlap with the electron-deficient cationic π orbital, resulting in the resonance-stabilized bridged silacycloprepenium ion. Evidence for the formation of such a reactive intermediate has been reported.[19]

21. Traylor, T. G.; Berwin, H. J.; Jerkunica, J.; Hall, M. L. *Pure Appl. Chem.* **1971**, *30*, 599.

22. The presence of a carbonium ion intermediate is inferred from a simple study of isomeric silanes **329** and **330** (shown below). Regardless of the ratio of starting disilanes, identical product ratios are obtained, and **xxvii** is the likely intermediate. See, Fleming, I.; Langley, J. A. *J. Chem. Soc., Perkin Trans. I.* **1981**, 1421.

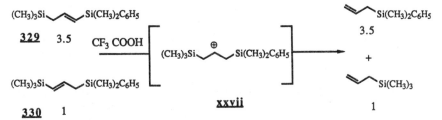

329 3.5

330 1   **xxvii**   1

23. Frainnet, E. *Bull. Soc. Chim. Fr.* **1959**, 1441.
24. (a) Shibasaki, M.; Fukasawa, H.; Ikegami, S. *Tetrahedron Lett.* **1983**, *24*, 3497. See, also (b) Koreeda, M.; Ciufolini, M. A. *J. Am. Chem. Soc.* **1982**, *104*, 2308.
25. (a) Fleming, I.; Au-Yeung, B.-W. *Tetrahedron* **1980**, *37* (Supp. #1), 13. (b) Au-Yeung, B.-W.; Fleming, I. *J. C. S. Chem. Commun.* **1977**, 79. (c) Fleming, I.; Pearce, A.; Snowden, R. L. *J. C. S. Chem. Commun.* **1976**, 182.
26. (a) Fleming, I.; Terrett, N. K. *J. Organomet. Chem.* **1984**, *264*, 99. (b) Fleming, I.; Marchi, D.; Patel, S. K. *J. C. S. Perkin I* **1981**, 2518. (c) Fleming, I.; Terrett, N. K. *Pure Appl. Chem.* **1983**, *55*, 1707. (d) Chow, H.-F.; Fleming, I. *Ibid* **1985**, *26* 397. (e) Fleming, I.; Terrett, N. K. *Tetrahedron Lett.* **1984**, *25*, 5103.
27. Hayashi and Kumada have established that acyclic allylsilanes preferentially react with various electrophiles with anti S_E' stereochemistry. See (a) Hayashi, T.; Konishi, H.; Ito, H.; Kumada, M. *J. Am. Chem. Soc.* **1982**, *104*, 4962, 4963. (b) Hayashi, T.; Ito, H.; Kumada, M. *Tetrahedron Lett.* **1982**, *23*, 4605. (c) Wickham, G.; Kitching, W. *J. Org. Chem.* **1983**, *48*, 612.
28. Hosomi, A.; Sakurai, H. *Chem. Lett.* **1981**, 85.
29. (a) Fleming, I.; Paterson, I. *Synthesis* **1979**, 446. (b) Paterson, I. *Tetrahedron Lett.* **1979**, 1519. (c) Reetz, M. T. *Angew. Chem., Int. Ed. Engl.* **1982**, *21*, 96. (d) Sadaki, T.; Usuki, A.; Ohno, M. *J. Org. Chem.* **1980**, *45*, 3559. (e) Sasaki, T.; Nakanishi, A.; Ohno, M. *Ibid.* **1981**, *46*, 5445. (f) Sasaki, T.; Nakanishi, A.; Ohno, M. *Ibid.* **1982**, *47*, 3219.
30. Hosomi, A.; Imai, T.; Endo, M.; Sakurai, H. *J. Organomet. Chem.* **1985**, *285*, 95.
31. Mohan, R.; Katzenellenbogen, J. A. *J. Org. Chem.* **1984**, *49*, 1238. See also ref. 29a.
32. Ohno, M.; Matsuoka, S.; Eguchi, S. *J. Org. Chem.* **1986**, *51*, 4553.
33. Sakurai, H.; Sasaki, K.; Hosomi, A. *Tetrahedron Lett.* **1981**, *22*, 745.
34. (a) Hosomi, A.; Endo, M.; Sakurai, H. *Chem. Lett.* **1978**, 499. (b) Nishiyama, H.; Itoh, K. *J. Org. Chem,* **1982**, *47*, 2498.
35. (a) Hosomi, A.; Sakata, Y.; Sakurai, H. *Tetrahedron Lett.* **1984**, *25*, 2383. (b) Hosami, A.; Sakata, Y.; Sakurai, H. *Carbohydr. Res.* **1987**, *171*, 223.
36. Danishefsky, S.; Kerwin, J. F., Jr. *J. Org. Chem.* **1982**, *47*, 3805.
37. Kozikowski, A. P.; Sorgi, K. L. *Tetrahedron Lett.* **1984**, *25*, 2085.
38. Recently, the preparation of C_1-allylated glycosides was used by Isobe *et al.* to construct a segment of okadaic acid. See Ichikawa, Y.; Isobe, M.; Goto, T. *Ibid.* **1984**, *25*, 5049.
39. (a) Bartlett, P. A.; Johnson, W. S.; Elliott, J. D. *J. Am. Chem. Soc.* **1983**, *105*, 2088. (b) Johnson, W. S.; Crackett, P. H.; Elliott, J. D.; Jagodzinski, J. J.; Lindell, S. D.; Natarajan, S. *Tetrahedron Lett.* **1984**, *25*, 3951.
40. Yamamoto, Y.; Nishii, S.; Yamada, J-i. *J. Am. Chem. Soc.* **1986**, *108*, 7116. See, also McNamara, J. M.; Kishi, Y. *J. Am. Chem. Soc.* **1982**, *104*, 7371.
41. (a) Calas, R.; Dunogues, J.; Deleris, G.; Piscioti, J. *J. Organomet. Chem.* **1974**, *69*, C15. (b) Deleris, G.; Dunogues, J.; Calas, R. *Ibid.* **1975**, *93*, 43.
42. Abel, R. W.; Rowley, R. J. *J. Organomet. Chem.* **1975**, *84*, 199.

43. Hosomi, A.; Sakurai, H. *Tetrahedron Lett.* **1976**, 1295.
44. Soai, K.; Ishizaki, M. *J. Chem. Soc., Chem. Commun.* **1984**, 1016.
45. Ojima, I.; Kumagai, M. *Chem. Lett.* **1978**, 575.
46. See (a) Bartlett, P. A. *Tetrahedron* **1980**, *36*, 3. (b) Yamamoto, Y.; Marvyana, K. *Heterocycles* **1982**, *18*, 357. (c) Hoffmann, R. W. *Angew. Chem. Int. Ed. Engl.* **1982**, *21*, 555.
47. For reviews of aldol stereoselection, see (a) Heathcock, C. H. In *Asymmetric Organic Reactions*, Morrison, J. D., ed., Academic Press, New York, 1984, Vol. 3, and references cited therein.
48. Heathcock, C. H.; Kiyooka, S.; Blumenkopf, T. A. *J. Org. Chem.* **1984**, *49*, 4214.
49. Yamamoto and co-workers have studied the same reaction using both allylsilanes and allylstannanes. They report that the stereoselectivity in these reactions with aldehydes is controlled by the choice of metal. See Yamamoto, Y.; Komatsu, T.; Maruyama, K. *J. Organomet. Chem.* **1985**, *285*, 31.
50. (a) Kiyooka, S.-i.; Heathcock, C. H. *Tetrahedron Lett.* **1983**, *24*, 4765. (b) Reetz, M. T.; Kesseler, K.; Schmidtberger, S.; Wenderoth, B.; Steinbach, R. *Angew. Chem. Int. Ed. Engl.* **1983**, *22*, 989.
51. (a) Fleming, I.; Kindon, N. D.; Sarkar, A. K. *Tetrahedron Lett.* **1987**, *28*, 5921. (b) Hosomi, A.; Hashimoto, H.; Sakurai, H. *Ibid.* **1980**, *21*, 951.
52. Pillot, J. P.; Dunogues, J.; Calas, R. *Tetrahedron Lett.* **1976**, 1871.
53. For an example of an S_E2' mechanism, see Sleezer, B.; Winstein, S.; Young, J. *J. Am. Chem. Soc.* **1963**, *85*, 1980.
54. Pillot, J. P.; Deleris, G.; Dunogues, J.; Calas, R. *J. Org. Chem.* **1979**, *44*, 3397.
55. (a) Hosomi, A.; Saito, M.; Sakurai, H. *Tetrahedron Lett.* **1980**, 3783. 5-Trimethylsilyl-1,3-pentadiene (57) has been treated with aldehydes and ketones, as well as acetals. See (b) Seyferth, D.; Pornet, J. *J. Org. Chem.* **1980**, *45*, 1721.
56. Fujisawa, T.; Kawashima, M.; Ando, S. *Tetrahedron Lett.* **1984**, 3213.
57. Santelli-Rouvier, C. *Tetrahedron Lett.* **1984**, *25*, 4371.
58. (a) Hosomi, A.; Hashimoto, H.; Sakurai, H. *J. Org. Chem.* **1978**, *43*, 2551. (b) Hosomi, A.; Hashimoto, H.; Kobayashi, H.; Sakurai, H. *Chem. Lett.* **1979**, 245.
59. (a) Hosomi, A.; Sakurai, H. *J. Am. Chem. Soc.* **1977**, *99*, 1673. (b) Sakurai, H.; Hosomi, A.; Hayashi, J. *Org. Synth.* **1984**, *62*, 86.
60. Recently, Hayashi and Mukaiyama have found that trityl perchlorate effectively promotes the conjugate addition of allylsilanes to α,β-unsaturated carbonyl compounds. See Hayashi, M.; Mukaiyama, T. *Chem. Lett.* **1987**, 289.
61. (a) Hua, D. H. *J. Am. Chem. Soc.* **1986**, *108*, 3835. (b) See also ref. 59.
62. Funk, R. L.; Bolton, G. L. *J. Org. Chem.* **1984**, *49*, 5021.
63. Heathcock, C. H.; Smith, K. M.; Blumenkopf, T. A. *J. Am. Chem. Soc.* **1986**, *108*, 5022.
64. Yanami, T.; Miyashita, M.; Yoshikoshi, A. *J. Org. Chem.* **1980**, *45*, 607.
65. Knapp, S.; O'Connor, U.; Mobilio, D. *Tetrahedron Lett.* **1980**, 4557.
66. Unpublished results of E. Logusch and G. Stork. For related work, see Stork, G.; Logusch, E. *Tetrahedron Lett.* **1979**, 3361.
67. (a) Blumenkopf, T. A.; Heathcock, C. H. *J. Am. Chem. Soc.* **1983**, *105*, 2354. See also (b) Mobilio, D.; DeLange, B. *Tetrahedron Lett.* **1987**, *28*, 1483.
68. (a) Majetich, G.; Casares, A. M.; Chapman, D.; Behnke, M. *Tetrahedron Lett.* **1983**, *24*, 1909. (b) Majetich, G.; Casares, A.; Chapman, D.; Behnke, M. *J. Org. Chem.* **1986**, *51*, 1745.
69. Hosomi, A.; Hashimoto, H.; Kobayashi, H.; Sakurai, H. *Chem. Lett.* **1979**, 245.
70. (a) Brownbridge, P. *Synthesis* **1983**, 1. (b) *Ibid.* **1983**, 85.
71. Pardo, R.; Zahra, J.-P.; Santelli, M. *Tetrahedron Lett.* **1979**, *47*, 4557.
72. House, H. O.; Sayer, T. S. B.; Yau, C. C. *J. Org. Chem.* **1978**, *43*, 2153.

73. Hosomi and Sakurai have made similar observations with other substrates; see Hosomi, A.; Sakurai, H. *Tetrahedron Lett.* **1980**, *21*, 955.

74. House, H. O.; Gaa, P. C.; VanDerveer, D. *J. Org. Chem.* **48**, 1661.

75. El-Abed, D.; Jellal, A.; Santelli, M. *Tetrahedron Lett.* **1984**, *25*, 1463.

76. Hosomi, A.; Sakurai, H. *Tetrahedron Lett.* **1977**, 4041.

77. (a) Naruta, Y.; Uno, H.; Maruyama, K. *Tetrahedron Lett.* **1981**, *22*, 5221. (b) Uno, H. *J. Org. Chem.* **1986**, *51*, 350.

78. Majetich, G.; Defauw, J.; Ringold, C. *J. Org. Chem.* **1988**, *53*, 50.

79. Nickisch, K.; Laurent, H. *Tetrahedron Lett.* **1988**, *29*, 1533.

80. For synthetic applications of 1,3-dithianes, see (a) Seebach, D. *Synthesis* **1969**, 17. (b) Seebach, D.; Corey, E. J. *J. Org. Chem.* **1975**, *40*, 231.

81. Westerlund, C. *Tetrahedron Lett.* **1982**, *23*, 4835.

82. Au-Yeung, B. W.; Fleming, I. *J. Chem. Soc., Chem. Commun.* **1977**, 81.

83. Deleris, G.; Dunoques, J.; Calas, R. *J. Organomet. Chem.* **1976**, *116*, C45.

84. Wada, M.; Shigehisa, T.; Akiba, K. *Tetrahedron Lett.* **1983**, 1711.

85. Ochiai, M.; Fujita, E. *Tetrahedron Lett.* **1983**, 777.

86. Fleming, I.; Paterson, I. *Synthesis*, **1979**, 446.

87. (a) Hayashi, T.; Kabeta, K.; Yamamoto, T.; Tamao, K.; Kumadda, M. *Tetrahedron Lett.* **1983**, *24*, 5661. (b) Carr, S. A.; Weber, W. P. *J. Org. Chem.* **1985**, *50*, 2782. (c) Oku, A.; Homoto, Y.; Harada, T. *Chem. Lett.* **1986**, 1495. See also (d) Tan, T. S.; Mather, A. N.; Procter, G.; Davidson, A. H. *J. Chem. Soc., Chem. Commun.* **1984**, 585.

88. Aratani, M.; Sawada, K.; Hashimoto, M. *Tetrahedron Lett.* **1982**, 3921.

89. For allylations of nitriles, see (a) Hamana, H.; Sugasawa, T. *Chem. Lett.* **1985**, 921. For the *in situ* generation and allylation of iminium ions, see (b) Larsen, S. D.; Grieco, P. A.; Fobare, W. F. *J. Am. Chem. Soc.* **1986**, *108*, 3512. For the photolysis of iminium salt–allylsilane systems, see (c) Ohga, K.; Mariano, P. S. *J. Am. Chem. Soc.* **1982**, *104*, 617. (d) Ohga, K.; Yoon, U. C.; Mariano, P. S. *J. Am. Chem. Soc.* **1984**, *49*, 213. For the iminium ion-mediated cyclizations of 4-aryl-1,4-dihydrophyridines, see (e) Saito, I.; Ikehira, H.; Matsuura, T. *J. Org. Chem.* **1986**, *51*, 5148. For the photochemistry of iodouracils, see (f) Hartman, G. D.; Phillips, B. T.; Halczenko, W.; Springer, J. P.; Hirschfield, J. *J. Org. Chem.* **1987**, *52*, 1136.

90. For the allylations of tricarbonylcyclohexadienyl–iron cations, see (a) Kelly, L. F.; Narula, A. S.; Birch, A. J. *Tetrahedron Lett.* **1980**, *21*, 2455. (b) Kelly, L. F.; Narula, A. S.; Birch, A. J. *Ibid* **1980**, *21*, 871. For the allylation of pyridinium ions, see (c) Kozikowski, A. P.; Park, P. *J. Org. Chem.* **1984**, *49*, 1674. For the allylation of acyl iminium ions, see (d) Kraus, G A.; Neuenschwander, K. *J. Chem. Soc., Chem. Commun.* **1982**, 134.

91. Sarkar, T. K.; Andersen, N. H. *Tetrahedron Lett.* **1978**, 3513.

92. Hosomi, A.; Shirahata, A.; Sakurai, H. *Tetrahedron Lett.* **1978**, 3043.

93. DePuy, C. H.; Bierbaum, M. M.; Flippin, L. A.; Grabowski, J. J.; King, G. K.; Schmitt, R. J.; Sullivan, S. A. *J. Am. Chem. Soc.* **1980**, *102*, 5012.

94. Majetich, G.; Hull, K.; Defauw, J.; Shawe, T. *Tetrahedron Lett.* **1985**, *26*, 2755.

95. The pentafluorosilicate ion is found in salts such as $[Ph_4AS] + [SiF_5]^-$. Data for SiF_5^- and also for similar species $RSiF_4^-$ and $R_2SiD_3^-$ are known. For examples of penta- or hexacoordinate silicon species see (a) Holmes, R. R.; Day, R. O.; Harland, J. J.; Holmes, J. M. *Organometallics* **1984**, *3*, 347. (b) Holmes, R. R. *J. Am. Chem. Soc.* **1978**, *100*, 433. (c) Somberg, D. J. *J. Organomet. Chem.* **1981**, *221*, 137. (d) Somberg, D. J.; Frebs, R. *Inorg. Chem.* **1984**, *23*, 1378; (e) Voronkov, M. G.; Deriglazov, N. M.; Brodskaya, E. I.; Kalistratova, E. E.; Gubanova, L. I. *J. Fluorine Chem.* **1982**, *19*, 299; (f) Corriu, R. J. P.; Guerin, C. *J. Organomet. Chem.* **1980**, *198*, 231; (g) Tamao, L.; Mishima, M.; Yoshida, J.; Takahashi, M.; Ishida, M.; Kumada, M. *J. Organomet. Chem.* **1982**, *225*, 151; (h) Voronkov, M. G. *Pure Appl. Chem.* **1966**, 13.

96. For a comprehensive review discussing hypervalent silicon species as intermediates in the reaction of orgnosilicon compounds see Corriu, R. J. P.; Cuerin, C.; Moreau, J. J. E., In *Topics in Stereochemistry*, Eliel, E. L.; Wilen, S. H.; Allinger, N. L., eds.; Wiley-Interscience, New York, 1984, Vol XV, pp. 43–198.

97. (a) Hosomi, A.; Kohra, S.; Sakurai, H. *Chem. Pharm. Bull.* **1987**, *35*, 2155. (b) Kira, M.; Kobayashi, M.; Sakurai, H. *Tetrahedron Lett.* **1987**, *28*, 4081. (c) Kira, M.; Sato, K.; Sakurai, H. *J. Am. Chem. Soc.* **1988**, *110*, 4599. (d) Hosomi, A.; Kohra, S.; Tominaga, Y. *J. Chem. Soc., Chem. Commun.* **1987**, 1517. (e) Hayashi, T.; Matsumoto, Y.; Kiyoi, T.; Ito, Y.; Kohra, S.; Tominaga, Y.; Hosomi, A. *Tetrahedron Lett.* **1988**, *29*, 5667.

98. As we speculated earlier, allylations that require only a catalytic amount of fluoride ion or employ fluoride ion catalysts that are only sparingly soluble in organic solvents (such as cesium fluoride) most likely involve pentacoordinate silicon nucleophiles. Here, we speculate that reactions that require a stoichiometric quantity of fluoride ion (or greater) most likely involve hexacoordinate silicon anions.

99. Unsymmetrically substituted allylsilanes have different substituents on the α- and γ-carbon atoms of the allylic system (the $SiMe_3$ group is by definition on the α-carbon).

100. For examples of substitution via a three-center interaction of an organometallic intermediate, see Negishi, E. In *Organometallics in Organic Synthesis*, Wiley-Intersceience, New York, 1980, Vol. 1, pp..67–73.

101. Hosomi, A.; Endo, M.; Sakurai, H. *Chem. Lett.* **1976**, 941.

102. (a) Ricci, A.; Degl'innocenti, A.; Fiorenza, M. *Tetrahedron Lett.* **1982**, *23*, 577. See also (b) Corriu, R. J. P.; Huynh, V.; Moreau, J. J. E. *J. Organomet. Chem.* **1983**, *259*, 283.

103. Our attempts to couple allyltrimethylsilane with allylic halides or benzylic halides using fluoride ion catalysis failed. Unpublished results of G. Majetich and A. M. Casares.

104. (a) Hosomi, A.; Sakurai, H.; Saito, M. T.; Sasaki, K.; Iguchi, H.; Sakaki, J.; Araki, Y. *Tetrahedron* **1983**, *39*, 883. 2-[(Trimethylsilyl)methyl]-1,3-butadiene has also been treated with acetals and acid chlorides. See (b) Hosomi, A.; Saito, M.; Sakurai, H. *Tetrahedron Lett.* **1979**, 429.

105. Corriu, R.; Escudie, N.; Guerin, C. *J. Organomet. Chem.* **1984**, *264*, 207.

106. Paquette, L. A.; Yan, T. H.; Wells, G. J. *J. Org. Chem.* **1984**, *49*, 3610.

107. Ricci, A.; Fiorenza, M.; Grifagni, M. A.; Bartolini, G. *Tetrahedron Lett.* **1982**, *23*, 5079.

108. Daviaud, G.; Miginiac, P. *Tetrahedron Lett.* **1973**, 3348.

109. Panek, J. S.; Sparks, M. A. *Tetrahedron Lett.* **1987**, *40*, 4649.

110. As with all intramolecular processes, the issues of regio- and stereoselectivity are governed by the ability or inability of the reactive centers to obtain a proper trajectory for reaction. A thorough discussion of this selectivity, however, exceeds the scope of this review.

111. Wilson, S. R.; Price, M. F. *Tetrahedron Lett.* **1983**, *24*, 569.

112. Kraus, G. A.; Hon, Y.-S. *J. Org. Chem.* **1985**, *50*, 4605.

113. Ipaktschi, J.; Lauterbach, G. *Angew. Chem. Int. Ed. Engl.* **1986**, *25*, 354.

114. (a) Nicholas, K.; Siegel, J. *J. Am. Chem. Soc.* **1985**, *107*, 4999. (b) Schreiber, S. L.; Sammakia, T.; Crowe, W. E. *J. Am. Chem. Soc.* **1986**, *108*, 3128.

115. (a) Johnson, W. S.; Newton, C.; Lindell, S. D. *Tetrahedron Lett.* **1986**, *27*, 6027. (b) Johnson, W. S.; Lindell, S. D.; Steele, J. *J. Am. Chem. Soc.* **1987**, *109*, 5852. Propargylsilanes have also been used as terminating groups in olefinic cyclization reactions. Despo, A. D.; Chiu, S. K.; Flood, T.; Peterson, P. E. *Ibid.* **1980**, *102*, 5120. (c) Schmid, R.; Huesmann, P. L.; Johnson, W. S. *Ibid* **1980**, *102*, 5122.

116. (a) Armstrong, R. J.; Harris, F. L.; Weiler, L. *Can. J. Chem.* **1982**, *60*, 673. (b) Armstrong, R. J.; Harris, F. L.; Weiler, L. *Can. J. Chem.* **1986**, *64*, 1002.

117. (a) Fleming, I.; Pearce, A.; Snowden, R. L. *J. C. S., Chem. Commun.* **1976**, 182. (b) Chow, H.-F.; Fleming, I. *J. Chem. Soc., Perkin Trans. I.* **1984**, 1815.

118. Johnson, W. S.; Chen, Y. Q.; Kellog, M. S. *Biol. Act. Princ. Nat. Prod.* **1984**, 55.

119. Lee, T. V.; Richardson, K. A.; Taylor, D. A. *Tetrahedron Lett.* **1986**, *27*, 5021.
120. (a) Wei, Z. Y.; Li, J. S.; Wang, D.; Chan, T. H. *Tetrahedron Lett.* **1987**, *28*, 3441. (b) Coppi, L.; Ricci, A.; Taddei, M. *J. Org. Chem.* **1988**, *53*, 911.
121. A comprehensive review entitled "The Stereochemistry of the Sakurai Reaction" written by Yamamoto and Sasaki is in press. This manuscript presents a thorough discussion of the diastereoselectivity observed in intramolecular addition of allylsilanes to simple carbonyl compounds and various unsaturated analogues.
122. Mikami, K.; Maeda, T.; Kishi, N.; Nakai, T. *Tetrahedron Lett.* **1984**, *25*, 5151.
123. Itoh, A.; Oshima, K.; Nozaki, H. *Tetrahedron Lett.* **1979**, *20*, 1783.
124. Nishitani, K.; Yamakawa, K. *Tetrahedron Lett.* **1987**, *28*, 655.
125. Denmark, S.; Weber, W. *Helv. Chim. Acta* **1983**, *66*, 1655.
126. (a) Trost, B. M.; Coppola, B. P. *J. Am. Chem. Soc.* **1982**, *104*, 6879. (b) Trost, B. M.; Hiemstra, H. *J. Am. Chem. Soc.* **1982**, *104*, 886.
127. Molander, G. A.; Andrews, S. W. *Tetrahedron Lett.* **1986**, *27*, 3115.
128. Cazes, B.; Colovray, V.; Gore, J. *Tetrahedron Lett.* **1988**, *29*, 627.
129. Trost, B. M.; Adams, B. R. *J. Am. Chem. Soc.* **1983**, *105*, 4849.
130. (a) Unpublished results of A. M. Casares. (b) Unpublished results of M. Behnke. (c) Unpublished results of S. Condon. (d) Unpublished results of K. Hull. (e) Unpublished results of D. Lowery. (f) Unpublished results of C. Ringold. (g) Unpublished results of S. Ahmad. (h) Unpublished results of J. Soria.
131. Wilson, S. R.; Price, M. F. *J. Am. Chem. Soc.* **1982**, *104*, 1124.
132. (a) Majetich, G.; Desmond, R.; Casares, A. M. *Tetrahedron Lett.* **1983**, *24*, 1913. (b) Majetich, G.; Desmond, R.; Soria, J. *J. Org. Chem.* **1986**, *51*, 1753. (c) Majetich, G.; Defauw, J.; Hull, K.; Shawe, T. *Tetrahedron Lett.* **1985**, *26*, 4711.
133. For a discussion of the diastereoselectivity observed in the intramolecular addition of allylsilane to enones, see Majetich, G.; Defauw, J.; Hull, K.; Lowery, D.; Shawe, T. Manuscript in preparation.
134. Majetich, G.; Defauw, J. *Tetrahedron* **1988**, *44*, 3833.
135. (a) Schinzer, D., *Angew. Chem. Int. Ed. Engl.* **1984**, *23*, 308; (b) Schinzer, D.; Solyom, S.; Bechker, M. *Tetrahedron Lett.* **1985**, *26*, 1831. (c) Schinzer, D.; Dettmer, G.; Ruppelt, N.; Solycm, S.; Steffen, J. *J. Org. Chem.* **1988**, *53*, 3823.
136. (a) Tokoroyama, T.; Tsukamoto, M.; Iio, H. *Tetrahedron Lett.* **1984**, *25*, 5067. (b) Tokoroyama, T.; Tsukamoto, M.; Asada, T.; Iio, H. *Ibid.* **1987**, *28*, 6645.
137. (a) Majetich, G.; Defauw, J.; Desmond, R. *Tetrahedron Lett.* **1985**, *26*, 2747. (b) Majetich, G.; Hull, K.; Desmond, R. *Ibid.* **1985**, *26*, 2751. (c) Majetich, G.; Behnke, M.; Hull, K, *J. Org. Chem.* **1985**, *50*, 3615. (d) Majetich, G.; Ringold, C. *Heterocycles* **1987**, *27*, 271. (e) Majetich, G.; Hull, K. *Tetrahedron* **1987**, *43*, 5621.
138. Our initial report of the intramolecular additions of allylsilanes to dienones occurred at the *Latest Trends in Organic Synthesis* conference held in Blacksburg, Va., May, 1984. Additional examples of this annulation strategy were reported at the 187th National Meeting of the American Chemical Society in St. Louis, Mo., April, 1984 and the 188th National Meeting of the American Chemical Society in Philadelphia, Pa., August, 1984.
139. All of the cyclizations in Chart 9 are intramolecular additions of an allylsilane moiety to a 3-vinylcycloalkenone. This description is far too general, yet formally derived names are impractical. In order to clarify this situation, we use the following convention: (1) the suffix "dienone" describes the 3-vinylcyclohexenone unit; (2) a locant for the allylsilane appendate is stated; (3) the nature of the allylsilane side chain is defined either as an *iso*-alkenyl or *n*-alkenyl substituent; and (4) geometric isomers or substitutions are ignored. Based on these conventions, substrates **260**, **262**, **264** and **265** are described as a 4-*iso*-butenyl-dienone, 4-*n*-pentenyl-dienone, a 5-*iso*-butenyl-dienone, and a 2-*iso*-butenyl-dienone, respectively.

isoalkenyl   n - alkenyl

140. For a comprehensive discussion of the diastereoselectivity observed in the intramolecular addition of allylsilane to dienones, see Majetich, G.; Lowery, D.; Hull, K. Manuscript in preparation.

141. Schinzer, D.; Steffen, J.; Solyom, S. *J. Chem. Soc., Chem. Commun.* **1986**, 829.

142. Tan, T. S.; Mather, A. N.; Procter, G.; Davidson, A. H. *J. Chem. Soc., Chem. Commun. 1984*, 585.

143. Wang, D.; Chan, T-H. *J. Chem. Soc., Chem. Commun.* **1984**, 1273.

144. (a) Armstrong, R. J.; Weiler, L. *Can. J. Chem. 61*, 214. (b) Armstrong, R. J.; Weiler, L. *Can. J. Chem.* **1986**, *64*, 584.

145. Cutting, I.; Parsons, P. J. *J. Chem. Soc., Chem. Commun.* **1983**, 1435.

146. Wada, M.; Shigehisa, T.; Akiba, K-y. *Tetrahedron Lett.* **1985**, *26*, 5191.

147. (a) Hiemstra, H.; Fortgens, H. P.; Speckamp, W. N. *Tetrahedron Lett.* **1985**, *26*, 3155. (b) Hiemstra, H.; Sno, M. H. A. M.; Vijn, R. J.; Speckamp, W. N. *J. Org. Chem.* **1985**, *50*, 4014. (c) Mooiweer, H. H.; Hiemstra, H.; Fortgens, H. P.; Speckamp, W. N. *Tetrahedron Lett.* **1987**, *28*, 3285.

148. (a) Gramain, J.-C.; Remuson, R. *Tetrahedron Lett.* **1985**, *26*, 327. (b) Gramain, J-C.; Remuson, R. *Tetrahedron Lett.* **1985**, *26*, 4083.

149. Grieco, P. A.; Fobare, W. F. *J. Chem. Soc., Chem. Commun.* **1987**, 185.

150. (a) Chen, S-f.; Ullrich, J. W.; Mariano, P. S. *J. Am. Chem. Soc.* **1983**, *105*, 6160. (b) Ahmed-Schofield, R.; Mariano, P. S. *J. Org. Chem.* **1985**, *50*, 5667. (c) Ahmed-Schofield, R.; Mariano, P. S. *J. Org. Chem.* **1987**, *52*, 1478. (d) Tu, S.-L.; Mariano, P. S. *J. Am. Chem. Soc.* **1987**, *109*, 5287.

151. Ochic, M.; Sumi, K.; Fujita, E.; Shiro, M. *Tetrahedron Lett.* **1982**, *23*, 5419.

152. Denmark is credited as having suggested the terms "synclinal" and "antiperiplanar" to define the orientations of the reactive centers prior to carbon–carbon bond formation.[125]

153. Trost, B. M.; Curran, D. P. *J. Am. Chem. Soc.* **1981**, *103*, 7380.

154. Trost, B. M.; Vincent, J. E. *J. Am. Chem. Soc.* **1980**, *102*, 5680.

155. (a) Hosomi, A.; Hoashi, K.; Kohra, S.; Tominaga, Y.; Otaka, K.; Sakurai, H. *J. Chem. Soc., Chem. Commun.* **1987**, 570. (b) Trost, B. M.; Otaka, K.; Shimizu, M. *J. Am. Chem. Soc.* **1982**, *102*, 4299.

156. (a) Galli, C.; Illuminati, G.; Manolini, L.; Tamorra, P. *J. Am. Chem. Soc.* **1977**, *99*, 2591. (b) Illuminati, G.; Mandolini, L.; Masci, B. *ibid.* **1975**, *97*, 4960.

157. (a) Baldwin, J. E. *J. Chem. Soc., Chem. Commun.* **1976**, 734. (b) Liotta, C. L.; Burgess, E. M.; Eberhardt, W. H. *J. Am. Chem. Soc.* **1984**, *106*, 4849.

158. Majetich, G.; Hull, K. *Tetrahedron Lett.* **1988**, *24*, 2773.

BIOGRAPHICAL SKETCHES
OF THE AUTHORS

Raymond Giguere is an Assistant Professor of Chemistry at Skidmore College, Saratoga Springs, New York. He received his doctoral degree in 1980 at the University of Hanover, West Germany, under the guidance of Professor H. M. R. Hoffman and was a research associate to Professor Harold Hart at Michigan State University before joining the faculty at Mercer University (1983–1988). His research is conducted at the undergraduate level with interests in the use of microwave heating in organic synthesis and the development of synthetic methodology involving intramolecular cycloadditions.

Tomas Hudlicky received his B.S. from Virginia Tech in 1973. He studied with Professor E. Wenkert at Rice University, where he received his Ph.D. in 1977. After a postdoctoral fellowship with Professor W. Oppolzer at the University of Geneva, he joined the faculty at Illinois Institute of Technology in Chicago. In 1982 he moved to Virginia Tech, where he is now a Professor of Chemistry. He received the A. P. Sloan Fellowship in 1981 and the NIH Research Career Development Award in 1984. His research interests include the development of enantioselective synthetic methodology, the design of new reactions, total synthesis of natural products, and microbial transformations of achiral hydrocarbons and the use of their metabolites in chiral synthesis.

George Majetich is an Associate Professor of Chemistry at the University of Georgia, Athens, Georgia. He received his doctoral degree in 1979 at the University of Pittsburgh under the guidance of Professor P. A. Grieco and

241

was a National Insititutes of Health Post-doctoral Fellow with Professor Gilbert Stork at Columbia University before joining the faculty at the University of Georgia in 1981. His research interests focus on the development of new annulation procedures using organosilicon chemistry.

Michael J. Taschner is an Associate Professor of Chemistry at the University of Akron. He received his B.S. degree in Chemistry in 1976 from the University of Wisconsin-Eau Claire and his Ph.D. in Organic Chemistry in 1980 from Iowa State University under the direction of Professor George A. Kraus. He worked for two years as an N.I.H. Post-doctoral Fellow with Professor Clayton H. Heathcock at the University of California-Berkeley before accepting his current position on the faculty at the University of Akron.

Advances in Molecular Modeling

Edited by
Dennis Liotta
Department of Chemistry, Emory University

"...as a result of the revolution in computer technology, both the hardware and the software required to do many types of molecular modeling have become readily accessible to most experimental chemists.
Because the field of molecular modeling is so diverse and is evolving so rapidly, we felt from the outset that it would be impossible to deal adequately with all its different facets in a single volume. Thus, we opted for a continuing series containing articles which are of a fundamental nature and which emphasize the interplay between computational and experimental results.*"*

—From the Preface

Volume 1, 1988, 213 pp. $78.50
ISBN 0-89232-871-1

CONTENTS: List of Contributors. Introduction to the Series: An Editor's Foreword, *Albert Padwa, Emory University.* Preface, *Dennis Liotta.* Theoretical Interpretations of Chemical Reactivity, *Gilles Klopman and Orest T. Macina, Case Western Reserve University.* Theory and Experiment in the Analysis of Reaction Mechanisms, *Barry K. Carpenter, Cornell University.* Barriers to Rotation Adjacent to Double Bonds, *Kenneth B. Wiberg, Yale University.* Proximity Effects on Organic Reactivity: Development of Force Fields from Quantum Chemical Calculations, and Applications to the Study of Organic Reaction Rates, *Andrea E. Dorigo and K.N. Houk, University of California, Los Angeles.* Organic Reactivity and Geometric Disposition, *F.M. Menger, Emory Universitry.*

JAI PRESS

Advances in
Metal-Organic Chemistry

Edited by
Lanny S. Liebskind
Department of Chemistry, Emory University

Organometallic chemistry is having a major impact on modern day chemistry in industry and academia. Within the last ten years, the use of transition metal based chemistry to perform reactions with significant potential in organic synthesis has come of age. *Advances in Metal-Organic Chemistry* will contain in-depth accounts of newly emerging synthetic organic methods that emphasize the unique attributes of transition metal chemistry problems in organic synthesis. Each issue contains six to eight articles by leading investigators in the field. Particular emphasis is placed on giving the reader a true feeling of the particular strengths and weaknesses of the new chemistry with ample experimental details for typical procedures.

Volume 1, 1989, 408 pp. $78.50
ISBN 0-89232-863-0

Advances in Oxygenated Processes

Edited by
Alfons L. Baumstark
Department of Chemistry, Georgia State University

Volume 1, 1989, 208 pp. $78.50
ISBN 0-89232-866-5

Volume 2, In preparation, Fall 1990
ISBN0-89232-950-5   Approx.$78.50

J A I P R E S S

Advances in
Solid-State Chemistry

Edited by
C.R.A. Catlow
Department of Chemistry, University of Keele

Advances in Solid-State Chemistry seeks to serve this expanding subject by producing timely and authoritative reviews on recent developments and techniques. In general, volumes in this series will include articles on materials, techniques and applications over a range of topics of particular contemporary interest; thematic volumes may also be published from time to time on specific topics or groups of related topics in rapidly emerging areas. The aim is to provide a balanced account of advances across the whole field of contemporary solid-state chemistry, for both the specialist and the wider communities of researchers and practitioners in the physical sciences.

Volume 1,1989, 333 pp.   **$78.50**
ISBN 0-89232-867-3